Verification of Arms Reductions

Verification of Arms Reductions

Nuclear, Conventional and Chemical

Edited by J. Altmann and J. Rotblat

In cooperation with the
Peace Research Institute Frankfurt (PRIF)

With 17 Figures

Springer-Verlag Berlin Heidelberg New York
London Paris Tokyo Hong Kong

Dr. Jürgen Altmann

Institut für Experimentalphysik,
Ruhr-Universität Bochum,
Postfach 102148,
D-4630 Bochum, FRG

Professor Dr. Joseph Rotblat

Pugwash, Flat A, Museum Mansions,
63A Great Russell Street, London
WC1B 3BJ, United Kingdom

ISBN-13: 978-3-642-46686-1 e-ISBN-13: 978-3-642-46684-7
DOI: 10.1007/978-3-642-46684-7

Softcover reprint of the hardcover 1st edition 1989

Preface

Our world is at a critical stage. Evolving global crises demand that mankind devote its resources to solving problems of the environment, of the developing countries, of inequalities within developed societies. Instead, the states of the world continue to increase their military spending; at present more than $1 trillion is spent annually on armaments, the lion's share coming from the nuclear powers and military alliances. However there are new signs that this pattern is changing. The INF Treaty between the USA and the USSR marks the first genuine, if only marginal, nuclear disarmament agreement. If this Treaty were followed by other agreements being negotiated at present, reductions of arms and military forces would ensue, and huge economic, technological and intellectual resources would be liberated for the important tasks facing mankind. The negotiations on the agenda are: a 50% reduction of strategic nuclear weapons, a ban of all nuclear tests, a ban of space weapons, drastic reductions of conventional forces, and a ban of chemical weapons.

Many factors have contributed to the recent improvement of the international climate. One major factor in concluding the INF Treaty has been the agreement on verification procedures, going far beyond anything achieved before in precision and detail. This Treaty may have broken the ground for other treaties: for most of the pending agreements, some outside presence in a national territory will have to be part of the verification package. What exactly will be needed and acceptable, is not yet fully defined. There are still questions to be answered, especially when dealing with detecting activities or objects that are small, mobile and easy to hide.

Bearing in mind the dangerous destabilization of a continuing arms race, taking into account the improved chances for international agreements, recognizing the crucial role verification plays in the conclusion of many treaties, and acknowledging the accelerating pace of verification research over the last few years, the Peace Research Institute Frankfurt decided to promote a "Workshop on Verification of Nuclear and Conventional Arms Reductions", held at London, UK, from Nov. 30 to Dec. 2, 1988, with some 110 scientists and government experts from many countries. This book is the collection of the written contributions to that workshop.

The Peace Research Institute Frankfurt and the organizing Committee wish to express their thanks to the John D. and Catherine T. MacArthur Foundation for a generous grant which made the Workshop possible.

The aim of the book is to present the experience gained in the verification of the most recent treaties; and to report the scientific state-of-the-art in those fields where agreements have not yet been concluded. We hope that it will contribute to further research in this rapidly developing field. In some cases, e.g. the Comprehensive Test Ban, verification methods have already been elaborated to impressing detail, it is other, usually political arguments that stand in the way of a treaty. There are, however, fields where the non-existence of established verification methods may prevent the conclusion of a treaty, or at least could be used as a convenient pretext for not concluding one. Such could be the case with verification of sea-launched cruise missiles, of stopping the production of fissile materials, of reductions of conventional arms, etc. There is a clear need for governments and independent scientists to enlarge activities in verification research and development. We are confident that this book will contribute to that end.

The book is divided into eight parts. Part I is devoted to a general discussion of verification and recent experience in bilateral disarmament. H. Müller (FRG) emphasizes the crucial role verification has to play in transforming the East-West conflict from confrontation to cooperation (Ch. 1). Experience with verification of the 1987 INF Treaty is reported by R. Summers (USA) and V. Koltunov (USSR), together with an outlook on strategic arms reductions (Chs. 2 and 3). In Ch. 4, G. Duffy (USA) discusses mechanisms for resolving compliance conflicts.

Part II deals with nuclear delivery systems, nuclear warheads and fissile material. V. Thomas (USA) looks into verification of sea-launched cruise missiles (Ch. 5). Possibilities for detecting nuclear warheads by the radiation emitted from the fissile materials are analysed in Ch. 6 by S. Fetter (USA), O. F. Prilutskii (USSR) and S. Rodionov (USSR). One possibility of disposing of materials from dismantled nuclear warheads, namely burning them in molten salt reactors, is proposed by A. Lecocq (France) in Ch. 7. The safeguards experience of the International Atomic Energy Agency is reviewed by D. A. V. Fischer (UK) in Ch. 8.

Nuclear weapons tests are the subject of Part III. T. Taylor (USA) discusses the role of nuclear testing for the development of nuclear weapons (Ch. 9). J. Leggett (UK) gives an overview of the recent developments for the verification of a nuclear test ban (Ch. 10). Some of these developments, namely joint international seismological experiments inside the USSR, are reported in more detail by R. A. Clark in Ch. 11.

Part IV deals with space weapons. Recent developments in the USA, especially the Strategic Defense Initiative, are reported by J. Pike (USA) (Ch. 12). S. Rodionov (USSR) discusses several possibilities for the verification of a ban on space weapons (Ch. 13).

The following three parts are devoted to conventional forces in Europe, beginning in Part V with discussions of conventional stability. This topic is the subject of A. Arbatov (USSR) and H. Siegmann (FRG) (Chs. 14 and 15).

Part VI reviews confidence and security-building measures in Europe. Experiences with manoeuvre observations, on-site inspections and future prospects are presented by a NATO and a WTO officer, B. A. Goetze (Canada) and M. Graczynski (GDR) (Chs. 16 and 17).

Part VII deals with the verification of conventional arms reductions in Europe. F. R. Cleminson (Canada) gives an overview over several different verification methods (Ch. 18). K. Jacob (FRG) discusses two methods specifically designed to allow verification of numerical limits on weapons systems while at the same time minimizing the information disclosure on military structures (Ch. 19). Remote sensing from satellites and aircraft is treated in detail by R. Reimers (FRG) in Ch. 20. J. Altmann (FRG) discusses possible uses of cooperatively emplaced short-range sensors to detect heavy military vehicles (Ch. 21).

Part VIII is devoted to verification of a chemical weapons ban. In Ch. 22, N. Smidovich (USSR) elaborates on the central role which mandatory challenge inspections will have to play. An overview over the structure of the verification system of the planned Chemical Weapons Convention is presented by J. P. Perry Robinson (UK) in Ch. 23.

We wish to acknowledge the help of the members of the Organizing Committee of the "Workshop on Verification of Nuclear and Conventional Arms Reductions", especially the London colleagues, who carried most of the organizational work. The Organizing Committee consisted of: Dr. Jürgen Altmann, Bochum/Frankfurt, FRG; Prof. Francesco Calogero, Rome, Italy; Dr. John Hassard, London, UK; Prof. Frank von Hippel, Princeton, USA; Prof. Sergei Kapitza, Moscow, USSR; Prof. Tom Kibble, London, UK; Dr. Patricia Lewis, London, UK; Dr. Harald Müller, Frankfurt, FRG; Prof. Joseph Rotblat, London, UK; Dr. Daniel Schertzer, Paris, France; Prof. Hartwig Spitzer, Hamburg, FRG.

In thanking all contributors to this volume, we also wish to acknowledge the help received in preparing the manuscript from: Wolfgang Baus, Elisabeth Grieger, Wolfgang Kaiser, Ilse Petry, Djundi Tjindra, Anke Steinbach, and Gudrun Weidner.

Bochum/Frankfurt,
London, July 1989

Jürgen Altmann,
Joseph Rotblat

Contents

Part I

Role of Verification and Recent Experience

1. Transforming the East–West Conflict: The Crucial Role of Verification

Harald Müller

1.1 Introduction

The politics of the East-West rivalry and arms control have rendered verification almost a political field of its own right. But this view is dangerously simplistic. Verification must be understood in its broader role in changing the character of the conflict. Conflicts are usually characterized by the variation of four distinct dimensions: the objects of conflict; the conflict instruments; the internal organization of the competing states for conducting the conflict; and the institutional relationships binding the competitors together. Verification affects the three latter of these four dimensions and has, thereby, an impact on conflict development that far surpasses its purported role in reinforcing the legal commitments of parties to particular agreements. Presently, we witness a unique opportunity to alter the basic tenets of the East-West relationship. In assessing the merits and prospects of verification, we must not lose sight of the crucial role verification might play in the overall transformation of the relationship between the Eastern and the Western world in the years to come.

Since the Napoleonic wars, states have pursued two competing roads towards the goal of preserving national security. On the one hand, they have continued to arm, and to resort to force if needed, to resolve conflicts and grant their territorial integrity: self-help in a highly anarchic world remained a justified and legitimate means of national security policy.

On the other hand, the devastating consequences of industrialized mass warfare, as it emerged, for the first time, in the long struggle after the French revolution, stimulated thoughts and efforts to avoid warfighting through cooperative arrangements between otherwise competing partners. Starting with the Vienna Congress and the Concert of Europe, these efforts grew larger with each breakdown of cooperation that led inevitably to a new round of ever more devastating warfare.[1]

The creation of the Red Cross and the refinement of the laws of warfare; the European peace movement, connected with the name of Bertha von Suttner and the Hague conferences before the first World War; the foundation of the league of nations, the naval limitation treaties, the considerations of offensive and defensive weapons, and the Briand-Kellogg Pact after the Great War, are all signals that increasingly, dissatisfaction with the traditional self-help system grew as the consequences of modern warfare became more cruel, and also more obvious.

Yet, it was not before the appearance of nuclear weapons that cooperative security, though still in a dangerously limited way, was translated into realistic operational concepts by governments in East and West, and it required the sobering shock of the Cuban missile crisis before these concepts bore their first fruits. Since then, each period of détente has produced more and deeper constraints on self-help, and each cooling-off period has ended with a reinvigorated effort to add to those constraints. Cooperative security has, not out of free will but of sheer necessity, conquered a role in national security policies in East and West, the continuing vigorous rivalry of self-help notwithstanding.[2]

1.2 Arms Control and Verification: Some Conceptual Thoughts

Classical arms control theory was developed in the late fifties and the early sixties in the United States, with some British contributions.[3] The prime purpose of arms control, according to these approaches, was to isolate and seal off the most dangerous and destabilizing areas of the nuclear arms race. It was fully understood that the conflict would continue. Islands of cooperation should hold their ground in a sea of persistent competition.[4]

Verification had to play an essential role in keeping the islands above the waters. For this purpose, verification was to be limited. Its reach was confined to its essential functions with relation to the few agreements that emerged. The further development of the instruments of self-help was sanctioned as legitimate. Hence, the boundaries of the insights into the military machineries of the rivals remained necessarily limited. Beyond the limits of treaties and agreements, concealment, deception and interference with national means of intelligence were quite justified. However, this created a strong temptation to push against those boundaries as far as possible, if such actions could foster an advantageous posture in the balance of power. This circumstance created a painful asymmetry in disfavour of the more open and transparent West, and made arms control in general and verification in particular an easy game of those who did not like cooperative security efforts in the first place.[5]

Verification is a highly subjective process. At each stage, from monitoring and surveillance, data processing, identification of points of political and strategic concern, evaluation of such disturbing findings, and decision on compliance policy, the subjective, political element gains in strength. There is ample room for all the evils contained in the pandora box of political misperception; and there is considerable decision stress, because in the security field, tolerable margins of error are small and only the worst-case analyst is assumed to stand always on the safe side.[6]

1.3 Verification and the Political Debate

Consequently, the same data led to extremely different opinions in the West, especially in the US discussion: those well disposed towards arms control assessed existing verification agreements as by and large satisfactory: they created useful insights, confirmed compliance, and strengthened the treaties and agreements concluded. For right-of-the centre politicians, and increasingly for many centrists, the continued competition changed the mindset by which verification data were interpreted. The more the Soviet Union was seen as ruthless and advantage-seeking, the less the confidence in the results and methods of verification. Assessing verification data became much an issue of the general East-West climate, on the one hand, and the balance of political power in the United States, on the other hand.[7]

The all-out attack of the first Reagan Administration against the existing verification regimes aimed at a strategy which I regard as basically flawed: the complete replacement of cooperation by the return to unfettered self-help competition as the single instrument of security policy. The policy apparently helped stimulate some critical review of previous policies in the Soviet Union, but conceptually, politically, and financially it was not sustainable. Nevertheless, some of the critical assessments of verification discussed in the American right have merits, though largely for reasons different from those promulgated by their authors.

The liberal defenders of arms control fought rightly for continuing a strategy of cooperative security. But their defence of verification achievements was partly based on weak grounds. It was quite natural and predictable that many people were bound to derive negative conclusions from the non-regulated areas of the East-West rivalry, and to interpret verification results accordingly. It is all right for analysts to divide a conflict into various functional areas and to search for solutions in each of those.[8] In the real political world, all these areas mix up in the minds of people. In a politician's brain, there is no Chinese wall between Soviet armament policy, verification, and geopolitical competition in the Horn of Africa, for example, as the Carter administration had to learn to its dismay. Even if we refuse a conscious policy of linkage on sound analytical and reasonable grounds, unconscious linkage is inevitable in the real world. Isolating arms control agreements from the other aspects of conflict, and understanding verification results as if they existed in a political vacuum, is a very weak spot in our defence of arms control. By the way, it was also a weak spot in the Soviet concept of détente in the Brezhnev era, and in the end, it did not work.

What has happened so far in arms control and verification affords an invaluable experience in the technical, political and legal fields. Not the least of these findings is that the strive for security cooperation in the nuclear age is so strong that a skeleton of cooperation survives even the heaviest storms of revived rivalry, as in the early eighties.[9]

But as a strategy the old concept of islands in a sea of competition is flawed, and the restricted role of verification was unsustainable. If arms control remains the only aspect of the rivalry which is regulated by cooperative restraint, and if even this aspect is confined to narrowly circumscribed areas, then we must expect a perpetual swinging forth and back of the pendulum of détente and confrontation. Verification will be subjected to alternating phases of overconfidence and unjustified distrust. It is thus high time to change our paradigm of security cooperation, and to give verification a new and more constructive and comprehensive place in it.

1.4 Towards Cooperative Security

It should be well understood that even in the new phase of East-West relations, self-help remains the residual resort of national security. Complete disarmament is not yet on the agenda. Each state will see to it that it retains a reliable and sufficient military instrument for deterrence and self-defence. Yet the great difference will be that the full spectrum of the East-West conflict, not only a highly selective set of modern weaponry, is to be submitted to agreed rules of mutual restraint. Three aspects merit particular attention in this context.

First, mutual restraint has to encompass the objects of rivalry, not only the means. Those who argue that limiting arms is useless because the roots of conflict lie elsewhere take their case too far, but they have got a point. Therefore, the increasing number of consultations on regional issues, and the solemn commitment to work towards regional settlements in crucial trouble spots are as important as arms control, and they will facilitate the task of arms control negotiations quite considerably. An increasing number of regional settlements will lead to a changed image of the East-West conflict: the relationship will be seen as less competitive in geopolitical terms. Consequently, the enemy will no longer be regarded as ruthlessly struggling for enhancing his power. In this context, the burden of verifying arms control agreement dwindles with the enemy image: more and more, verification will turn towards confirming the sincerity of the other side's commitment to self-restraint, and it will no longer be suspected as being a dangerously deceptive cover for power expansion. At the very least, the claims of those continuing stubbornly to promulgate this assertion will find fewer and fewer believers.

Secondly, security cooperation must encompass as many fields of military activity and armament as possible. Ideally, it should be all-comprehensive. The more areas of the arms competition are contained or even sealed off by mutual agreements, the less temptation will there be to undo or to circumvent those agreements through exploiting loopholes in non-regulated areas. Where grey areas continue to exist - and this will be inevitable for some time to come - the rule of behaviour must be self-restraint rather than exploitation of loopholes and ambiguities. Both sides have to learn a lot in this respect. If any proof was needed that such exploitation desires exist in the West as well, the debate on the "extensive" interpretation of the ABM

Treaty and the study on "Discriminate Deterrence" were convincing evidence. Where one-side advantages exist, they must not be pressed to their limits, but preferably be traded in for reasonable concessions.[10]

Under these circumstances, the role and understanding of verification will be subjected to a landmark change. If the hot pursuit of circumvention, rechanneling and cheating opportunities is discontinued, verification results will meet a far more relaxed audience everywhere. Marginal errors, inaccuracies and uncertainties will be seen as less risky. People will be more inclined to judge verification for its own merits and standards rather than scrutinize it for loopholes and alleged weaknesses which are derived from the most unrealistic scenarios.

Thirdly, security cooperation must lead to the restructuring of the military-industrial-political machines on both sides. The mere existence of such complexes with vested interests on both sides does much to maintain suspicion and anxiety. The societal roots of military activities have engendered complex theories in East and West about the structural and systemic causes of the existing threat. If those circles responsible for maintaining self-help instruments on either side are assessed as honest and reliable, and no longer as genuinely aggressive and ruthless, then verification will have an easier time; with no suspected masterminds behind deception and circumvention, the confidence-building purpose of verification can be realized efficiently.

1.5 The Role of Verification in the Process of Cooperative Security

The development of cooperative security will be a long process. However bold the initiatives emanating from Moscow are, the structure of the world will change in steps rather than in jumps. In this protracted process, verification has to fulfill four essential functions.

In the first place, verification will continue to play its proper, direct role: confirming compliance with existing and emerging arms control and arms reduction agreements. For this purpose, adequacy remains the valid and reasonable standard: verification must grant a high probability of detection for all strategically significant violations of arms control commitments.[11] To carry the standard further *conceptually* means to request the impossible: in a world where human mind makes the last judgment, margins of error and uncertainty are inescapable.

Yet the role of verification does not stop here. If we construct verification with this function in mind only, we will fail on political grounds, even in a benevolent environment. It was the error of liberal arms controllers, but also of the Soviet Union, in the first phase of arms control, that a strictly rational-functional construction of verification measures could neglect political side-effects. They understood properly that competition would go on in other areas, and precisely for this reason they wanted to restrict verification strictly to its specific role within arms control agreements. The Soviets naturally wished to keep their undeniable advantage which their secrecy afforded them in the strategic competition. Legally, this position was impeccable, but politically it was badly flawed. The right-wing critics of arms control could expose this failure mercilessly in their campaign against cooperative security at large. Whatever our conceptual understanding of verification is, in the real world conclusions will be drawn from experiences in unregulated areas, and verification results will be reinterpreted in the light of these experiences. And vice versa, ambiguous behaviour in third areas will be assessed in the light of already tarnished experiences with verification. The perceptual osmosis between regulated and unregulated areas leads inevitably to a vicious circle of distrust and hostile behaviour which, as a self-fulfilling prophecy, must unravel cooperative security in the end. For this reason, we must include in our concept of verification three additional functions.

Secondly, by accepting far-reaching verification obligations beyond what is strictly needed for verifying a specific agreement, the parties create a unique opportunity to demonstrate to each other their political preference for security cooperation over self-help cooperation:

- They give priority to transparency over secrecy, if a conflict between reducing ambiguity contra protecting aspects of their military posture arises. In this way, they communicate to each other the essential message that they value security cooperation higher than a marginal addition to their own strength. In some cases, this may impose restrictions on preferred ways of deployment and operations. It all depends where priorities are seen. The verifiability of an arms control measure is the result of both the appreciation of the strategic option foregone and the intrinsic difficulty of a weapon to be verified. By raising the verification effort, the latter barrier can presumably always be overcome. But this effort will not be made if pursuing the strategic option is preferred to limiting it. It is this constellation, not intrinsic unverifiability, which impedes arms control for sea-launched cruise missiles, or a comprehensive test ban.
- Moreover, they prove their readiness to incur even serious changes in their domestic structures in order to eliminate elements that hamper cooperation. For the Soviet Union, this means bold steps away from its traditional secrecy and frightening concentration of state power, towards opening the thoroughly protected bits and pieces of its defence and military fabric. It is encouraging to see the first steps taken on this long and stony road, but much remains to be done.

For the West, mainly for the US military and intelligence establishment, the requirements are obviously more challenging than we were inclined to believe in the past, when Soviet stonewalling saved the West from seriously thinking through what openness really means in military terms. The INF Treaty and current chemical arms control talks have revealed that effective verification is no cost-free undertaking in market-economy democracies either:

- The West has to admit a degree of intrusion into the property rights of private industry that stimulates strong reservations and objections in the companies concerned, and requires strong political determination for arms control considerations to prevail.
- The West may have to compromise its desire to maintain as large a technological advantage over the East as possible. Highly intrusive inspections may have the side effect of giving away some technical information here and there which we may otherwise prefer to protect. However, this loss may be well worth the candle if the *overall process* enhances our security.

Thirdly, more liberally formulated verification schemes will inevitably lead to a growth in collateral information conveyed to the verifying party. The technical insights just mentioned are only part of this story. More important is political and social information which will be learned by inspectors and, through them, by the political systems on both sides. Collateral information gathering was a bone of contention that has burdened arms control negotiations in the past. However, if our goal is a system of cooperative security, such incidental gains in transparency should be seen not as a liability but as a blessing. Tolerance of such collateral information gathering conveys a sense of trust in the adversary and, by itself, one's own peaceful intentions: no party poised to prepare a premeditated attack is likely to accept any leak of information to its adversary beyond its narrowest legal obligations. Moreover, acquisition of collateral information across the broad spectrum of arms control will most probably help to reduce ambiguities by mutual support of evidence between different agreements. Reciprocity, of course, is the precondition for such collateral information gathering to work satisfactorily.

Lastly, there is the gain in human contacts among those parts of either society which need mutual recognition and appreciation most: the military. Verification schemes bring together people who are supposed to fight and to kill each other if it comes to it. The very few friendly encounters caused by the Stockholm Agreement and the INF Treaty have already made an impact on those officers representing their countries at manoeuvres, at INF destruction sites and the like. The unlearning of enemy images by those most likely and inclined to maintain worst-case assessments is an invaluable side-effect of verification. This effect should be fully appreciated, and future verification schemes should be designed with this effect in mind.

The configuration of these four functions makes one conclusion immediately evident: in the old days, the assessment of hostile military capabilities has led directly to judgments of enemy intentions. Peace researchers have fought this simple logic often enough, but with limited success only. In a world where most international relations are characterized by the race for power, and where many activities are not transparent, it is quite reasonable to derive conclusions about your enemy's plans from the only hard intelligence you can get: capabilities. In the world of security cooperation, in contrast, verification measures designed to provide information about capabilities will also convey information about intentions. This is the purpose of the second, third, and last functions of verification. And through this enhancement of mutual trust verification will greatly contribute to the transformation of the East-West conflict.

It is for this purpose, and not for gaining court-proof evidence that I believe future arms control agreement should contain extensive verification rules rather than a narrowly circumscribed approach.

1.6 Methods of Verification

National technical means of verification (NTM) must be developed further. They provide the firm ground from which all other methods extend. As we start from a situation in which security cooperation is not well developed, only solid NTM technology will create the self-confidence that permits nation-states to enter more far-reaching cooperative endeavours.[12]

The dual-purpose character of many NTM technologies, i.e. their capabilities to acquire war fighting-essential data, is worrisome but will lose importance once the network of security cooperation grows more elaborate.[13]

At this point, the exchange of monitoring technology might be seriously considered. Such an exchange was practiced during the nuclear test calibrating exercises; it was also conceived, though never realized, in the PNE (Peaceful Nuclear Explosion) Treaty, and it is implicit in the internationalization of safeguards equipment technology in the IAEA. The additional confidence-building effect of such an exchange hardly needs mentioning (see below).

Cooperative measures should be expanded as far as possible, if not for technical, then for politico-psychological reasons. Such measures facilitate the observation of national strategic assets by the adversary. Hence, they convey the feeling that there is little to hide indeed. Moreover, they have an important educational and training effect on the respective armed forces. The incorporation of cooperative measures in the organizational routines of the military changes the daily life of soldiers for the sake of security cooperation. The importance of this effect on the thinking of military persons should not be underrated.

Thirdly, on-site inspection. Liberal defenders of arms control have tended to play down their role. Routine inspections they accept, but they see little merit in on-challenge on-site inspections. Little new information is to be gained, it is argued, and a violator of arms control agreements would not permit on-site collection of evidence anyway; in politically tense times, on-site inspections could be abused by the enemies of arms control, and ambiguous evidence would be misinterpreted with or without on-site inspection.[14]

I respect the thrust of the argument but disagree with its substance. Again, too narrow an understanding of arms control verification underlies this position. My suspicion is that the main reservation against on-site inspections stems from the cumbersome experience in previous negotiations with the Soviet Union; then, it proved immensely difficult to persuade the suspicious and secretive Soviets to accept what they regarded as thinly veiled espionage. The enormous difficulties involved, I suppose, still lead arms controllers to discard or devaluate this option though the Soviet attitude has changed so much. This change advises a reconsideration of the issue; the discovery of systemic difficulties with on-site inspections on the Western side promises that negotiations on this topic will be more balanced and symmetric in the future. Such symmetry augurs better chances to achieve balanced compromises.

In principle, there are several reasons why on-challenge on-site inspections, in all likelihood, will and should remain an inevitable part of the verification picture in a transition towards cooperative security.

The first reason is of politico-psychological nature. After all, human beings believe most in what they can see and touch; on-site inspection has an incomparable psychological appeal that goes far beyond its factual contribution to a verification system. The sheer possibility to go somewhere and to look if a suspicion arises will possibly strengthen confidence without a single such inspection being conducted. This appeal cannot be matched by a hundred highly precise satellite-pictures.

Secondly, in accepting on-challenge on-site inspections a state gives the most far-reaching demonstration of its innocence; this is a political corollary to the psychological effect just described. No country poised to prepare a surprise attack would admit foreigners to roam around while its forces were gearing up for a ruthless assault; neither would anybody preparing a large-scale break-out, or a militarily significant creep-out of arms control obligations, submit its activities to the eyes of suspicious observers.

Some liberals have insisted that cheaters would never admit foreign inspectors and, thus, on-challenge verification would not be a useful instrument of verification. But the time-honoured and correct liberal counter argument against right-wing arms-control criticism was that verification is not for collecting court-proof evidence. Refusing inspectors, or treating them in the way Ghaddafi did to the journalists during the "inspection" of the Rabta chemical complex, would be overwhelming evidence that something is wrong. Such an extraordinary event would provide ample justification for a proportionate reaction. The absence of outright refusal to admit inspectors, or of unusual interference with their activities, would thus strengthen the mutual confidence between parties, the very political purpose verification is constructed for in a conflict-transformation strategy.

There is some uneasiness that requests for on-challenge inspections in themselves could precipitate a deterioration of mutual relations through the open display of unusual suspicion. This may be true after a long absence of any inspection requests. In this case, asking for the admittance of inspectors would indeed mean throwing stones in still waters. The danger, however, can be prevented by applying the reasonable practice developed in the CDE agreement context and also envisaged for INF: the quota for on-challenge inspection is being routinely used.[15] Not only does this help to train inspectors in their new trade; by accustomizing publics, governments and military establishments to on-challenge inspections as a normal activity, the sensational touch of such exercises is eliminated. If access is needed once for clarifying a truly ambiguous situation, this can be done in the calm and sober atmosphere of normality.

Lastly, some fear that unjustified requests for frequent on-challenge on-site inspections could be abused by enemies of arms control in order to poison the atmosphere between East and West, and to collect ambiguous evidence to bolster fabricated claims of cheating. In many cases, evidence gained by inspectors on the spot may not be completely clear and subject to different interpretations. Arms control critics could then build a case for destroying an agreement by one-sided presentations of alleged on-site inspection findings. There is certainly a point here. The late seventies and early eighties present impressive proof that real and forged evidence can be combined to construct a considerable, politically effective case for abandoning arms control. This was, one remembers, in a verification environment without any on-challenge on-site inspections. The same could possibly be done in a verification system characterized by dense inspection schemes. However, inspection would hardly make things worse; if the overall state of a relationship had already deteriorated to a level where such a contrived denunciation could find many believers at home, the situation was presumably hopeless anyway, and the abuse of on-site inspection would deliver a coup de grace rather than the initial bombardment softening the defences of arms control. If, on the other hand, the situation was not decisively bound to get worse, and if the partner was truly innocent, then on-site inspection would present better opportunities for arms control proponents to build a case for refuting the accusations than in the absence of such inspection opportunities. More important, long experience with a well-entrenched inspection scheme is more likely to prevent the emergence of a political atmosphere in which dubious allegations of arms control cheating can succeed in stimulating, enhancing, and finally harnessing public suspicion. Hence, rather than presenting a particularly virulent problem of affording ammunition to arms control opponents, far-reaching on-site inspection may simultaneously contribute to munition for the other side as well, and maybe even better munition at that. And, from the beginning, it will prepare the political battlefield in a way that would strengthen the defensive positions of those devoted to arms control.

As an aside, on-challenge inspections provide a further field for broadening the acquaintance of the officer corps with their counterparts on the other side. It is thus part of the seminal education and socialization process analysed above, and has some additional worthiness in that regard.

To summarize, rather than being a panacea for all the ills of arms control, on-challenge on-site inspection is a useful supplementary instrument within a broader context. It is certainly not without defects and dangers, as the more thoughtful observers were right to suggest. Yet, within the overall political context in which this far-reaching incursion into the sacred realm of national sovereignty is only conceivable, namely one of reduced mutual hostility and strengthened ties, its positive consequences are most probable to overshadow the residual dangers. For this reason, a prudently devised on-challenge inspection scheme should be an integral part of arms control verification.

1.7 Technology and Verification

It is not the intention of this brief paragraph to summarize the complex relationship indicated in its headline; other contributions in this volume by more qualified authors respond to this callenge exhaustively. Here, the purpose is to indicate in a few sentences which aspects are important from the perspective of conflict transformation towards cooperative security. Four such aspects can be identified.

1. The heavy emphasis which this paper puts on the political and psychological aspects of verification should by no means devaluate the significance of the technological means of proper verification. On the contrary: a correct understanding of the interplay between psychology, politics, and technology points clearly to the critical contribution that verification techniques make to long-term confidence-building between enemies. Uncompromised trust in the own national means of verification is the first and foremost building block on which successful verification must rest. This is so because national means link the cooperative venture of arms control with the traditional way of security policy, self-help. Through national means, arms control rests safely on the residual means of self-help, and vice versa.[16] If the verification building relies more and more on other blocks as well, all the better. But the fundamental meaning of the solid first building block must not be forgotten. If everything goes wrong, here is a fallback position on the way of retreat to a self-help security policy. Without this certainty, nation states would never commit themselves to the tricky path of security cooperation.
2. In the future, it will be urgent to understand better the interplay between technical and non-technical means.[17]
   We have to evaluate comprehensively the opportunities of networking the collateral findings of agreement-related technical and non-technical means in order to enhance confidence among the parties. Rather than to isolate different agreements hermetically from each other, it should be accepted as legitimate to exchange information among them (e.g. satellite intelligence should legitimately be discussed by the board of governors of the IAEA).

3. The sharing, and, eventually, internationalization of verification technology should be considered a serious option. Intelligence communities shy away from this suggestion, understandably so, as the dual-purpose mission of many verification devices is well-known, and other technology applied in the verification field nationally may contain sub-systems and components well applicable to the military task. To give away such assets is particularly difficult for the West, first and foremost for the United States, where technological superiority continues to be seen as a legitimate compensation for advantages of the Soviet Union in numbers and geography. Yet, after the breathtaking offer of former President Reagan to share SDI technology with the other superpower, such a possibility should not be a taboo anymore.
   That the mutual application of state-of-the-art verification technology is advantageous for confidence-building is obvious. A legitimate objection is that knowing the technique of the enemy presents new opportunities for countermeasures, and indicates the roads to be taken for camouflage and deception. This, however, can be counteracted by supplementary agreements on cooperative measures. Even the SALT agreements contained such precautions, though still in a rudimentary way. The INF agreement and, more so, the preliminary agreements reached in START, go some steps farther. It is quite conceivable that all exchanges of verification technologies are formalized in agreements prescribing a set of cooperative measures aimed at preventing the abuse of the freshly gained knowledge of the partner's technical verification capabilities.
4. Finally, a rather curious thought: it appears that the purposeful development of verification-specific technologies - as opposed to the development of such means as mere side-effects of primarily intelligence-gathering activities - may become a growth market in the future. Of course, in the International Atomic Energy Agency, safeguards-specialized technologies have been existing for a while, and a well-funded safeguards support programme in several advanced countries continues to push the front of safeguarding techniques forward. But there are prospects for an expansion of this "industry": for example, the extensive utilization of tagging devices such as proposed by the WEU for conventional arms control, or by others for cruise missiles, would expand this market considerably;[18] further possibilites can be seen as one reads the contributions to this volume. This development is significant: it expresses symbolically the attention paid to security cooperation in monetary terms. A minor but distinguishable part of military budgets, RD&D (Research, Development & Demonstration) activities and, finally, hardware production would be redirected towards arms control purposes. This would be a healthy evolution, particularly in Western countries: vested interests in laboratories and even in industry would emerge supporting arms control endeavours rather than armament projects. Modest as those interests would be compared to the existing stakes in military hardware, they would be politically significant, given

their total absence under present circumstances. Is there a more efficient counterweight conceivable to the "military industrial complex" than a properly lobbying "arms control industrial complex"?

1.8 Conclusions

The process towards security cooperation is a protracted one, and a system of cooperative security is a long way down the road. Armament and unilateral strategies will be with us for the foreseeable future. Arms controllers have often complained about the verification catch-22: verification is supposed to build trust among enemies, but to function in this way, trust on both sides must exist in the first place. A quantum jump may be needed to overcome this vicious circle.[19] The present changes in the Soviet Union may well provide this quantum jump. The nuclear age has forced upon very reluctant great powers the inescapable need to cooperate for the sake of their security and their very survival. Verification is part and parcel of this cooperation, it is indeed indispensable. The present window of opportunity for security cooperation may, for the first time in history, afford verification the chance to fulfill its role for maintaining cooperation and for building long-lasting confidence. This great chance encompasses all the important aspects: legal, technical as well as psychological and political. The chance must not be missed.

1.9 Notes and References

1 Robert Jervis, Security Regimes, in: Stephen D. Krasner (ed.), International Regimes, Ithaca NY: Cornell University Press, 1983, pp. 173-194

2 I have pursued this thought further in my paper "Selbsthilfe oder Kooperation? Regimes in der Sicherheitspolitik". Paper, presented to the Biennial Conference of the German Society for Political Science, Darmstadt, September 1988

3 The best-known classic is: Donald G. Brennan (ed.), Arms Control, Disarmament, and National Security, New York: Brazilla, 1961

4 This picture is adapted from: Joseph S. Nye, Nuclear Learning and U.S.-Soviet Security Regimes, in: International Organization, vol. 41, no. 3, Summer 1987

5 See the summary of: Richard N. Haass, Verification and Compliance, in: Albert Carnesale, Richard N. Haass, Superpower Arms Control, Setting the Record Straight, Cambridge MA: Ballinger, 1987, pp. 303-328; for an example of the conservative assault see: Malcolm Wallop, Angelo Codevilla, The Arms Control Delusion, San Francisco: Institute for Contemporary Studies, 1987

6 The reasons are well explained by: Robert Jervis, Cooperation under the Security Dilemma, World Politics, vol. 30, pp. 167-214, January 1978

7 Michael Krepon, The Politics of Treaty Verification and Compliance, in: Kosta Tsipis, David W. Hafemeister, Penny Janeway (eds.), Arms Control Verification - The Technologies That Make It Possible, Washington etc.: Pergamon Brassey's, 1986, pp. 20-32

8 This technique is eloquently defended by: Manfred Efinger, Volker Rittberger, Michael Zürn, Internationale Regime in den Ost-West-Beziehungen, Frankfurt: Haag und Herchen, 1988

9 Alexander L. George, Philip J. Farley, Alexander Dallin (eds.), U.S.-Soviet Security Cooperation - Achievements, Failures, Lessons, New York/Oxford: Oxford University Press, 1988

10 The Soviet record is tracked by: Gloria Duffy, The Sources of Soviet Compliance Behavior, in: Michael Krepon, Mary Umberger (eds.), Verification and Compliance, A Problem-Solving Approach, London: Macmillan Press, 1988; for the political consequences see: Krepon (note 7) and: Michael Krepon, The Political Dynamics of Verification and Compliance, in: William C. Potter (ed.), Verification and Arms Control, Lexington MA: Lexington, 1985, pp. 135-152

11 For a concise discussion of the adequacy concept see: Ralph Earle II, Verification Issues from the Point of View of the Negotiator, in: Tsipis et al. (note 7), pp. 14-20

12 Noel Gaylor, Verification, Compliance, and the Intelligence Process, in: Tsipis et al. (note 7), pp. 3-7

13 William F. Rowell, Arms Control Verification - A Guide to Policy Issues for the 1980s, Cambridge MA: Ballinger, 1986, pp. 51-58

14 See e.g.: James A. Schear, Cooperative Measures of Verification: How Necessary? How Effective? pp. 7-36, and Mark M. Loventhal, Joel S. Wit, The Politics of Verification, pp. 153-168, in: Potter (note 10); Allan S. Krass, Verification - How Much is Enough?, London: Taylor and Francis, 1985, pp. 212-222; a measured assessment of OSI coming to slightly more sceptical results than the present author is: Michael Krepon, Sidney Greybeal, On-Site Inspections, in: Krepon, Umberger (note 10), pp. 92-108

15 Insights of an On-Site Inspector, Arms Control Today, 4 November 1988

16 So Gaylor (note 12)

17 For a good discussion, see: Krass (note 14), Ch. 4

18 Western European Union, Requirements of a Tag for Conventional Arms Control, AG I (88) D/10, Paris: WEU, 1988; Richard L. Garwin, Tags and Seals for Verification, The Council of Arms Control Bulletin, no. 40, pp. 3-4, October 1988; John A. Adam, Verification: Special Sensors, Seals, Tags, Pulses, International Herald Tribune, p. 4, 26 Jan. 1988

19 Krass (note 14), pp. 166-167

2. Experience with INF Treaty Verification and Prospects for Effective Verification of Strategic Arms Reductions

Robert A. Summers[1]

2.1 Introduction

The term verification conjures up many different concepts from both technical and political perspectives. In the not too distant past, some perceptions had been that it was a convenient tool to prevent the successful negotiation and implementation of arms control treaties. Other perceptions have included idealized concepts of "perfect" verification. But perfect verification - 100% assurance that the other side has lived up to all provisions of an agreement - is not achievable, and, indeed, is not necessary in order to achieve and implement viable arms control and arms reductions agreements. What is needed is: 1. a reasonably high level of confidence that the other party is living up to its commitments; 2. a reasonably high level of deterrence against non-compliant behaviour; and 3. a timely indication if the other party should not live up to its commitments.

Achieving these objectives requires, first of all, the careful drafting of treaty provisions with verification needs in mind. In particular, treaty limitations must be defined - to the greatest extent possible - in terms of characteristics or features which can be detected by the verifying party. In past treaties, this has meant dependence on one's own national technical means of verification (NTM). However, advancing technology - small weapon size and mobility - has made this single approach very difficult. The advent of operational land-mobile missiles is a useful example. The inherent limitations of NTM makes sole dependence on NTM for verification infeasible.

However, careful and intensive study of this problem revealed that there were schemes involving a combination of NTM with cooperative measures on the ground which could achieve effective levels of verifiability for land-mobile ballistic missile systems. These concepts - together with some striking shifts in political winds - enabled the negotiation, signing, and implementation of the INF Treaty which provides for the complete elimination of an entire class of nuclear ballistic missile systems - those with ranges between 500 and 5,500 km - within a period of three years.

This Treaty was signed on December 8, 1987 and entered into force on June 1, 1988. Thus, we now have about six months experience with implementation of the Treaty. In order to examine our experience with this implementation, it is important to describe the details of this lengthy treaty and its complex verification regime. As one of my esteemed senior colleagues has said, "the devil is in the details." So we have to start there.

2.2 The INF Verification Regime

2.2.1 The Overall Concept

Verification is not an adjunct or afterthought to this Treaty. Rather, the verification concept is woven throughout virtually every article of the Treaty. This is because the USA designed its entire treaty draft with verification in mind, and insisted on negotiating verification aspects in parallel with, or as a part of, negotiating other aspects. Many of these verification aspects touched on sensitive issues on both sides, so that some decisions were not made until the final stages of negotiations.

The INF Treaty regime is designed both to control the declared *overt* INF inventory and to make as complicated and costly as possible the acquisition of any *covert* illegal inventory. The bulk of the treaty language focuses on the declared overt inventory.

The declared overt inventory is controlled by specific geographical and movement constraints, and is confirmed by on-site inspection (OSI). The geographical constraints are based on a partitioning of the Soviet Union into limited areas where INF missiles systems are allowed, and a larger residual area where they are prohibited unless in announced transit. In this context, we see the principal function of National Technical Means as not to "count" those declared INF systems, but, rather, to detect any illegal systems.

Destruction of the INF systems is observed and confirmed by US inspectors.

Covert forces could derive from either: 1. systems prodced before the Treaty in excess of the declared baseline inventory; or 2. illegal systems produced covertly during or after the period of elimination. Maintenance of a *militarily useful* covert force is rendered difficult by the immediate flight test ban together with NTM search for illegal systems. The immediate flight test ban will render extremely risky flight testing, which is important for establishing the reliability of systems produced even from existing designs. Also, both NTM monitoring and the flight test ban will make very risky the exercising of militarily significant INF forces after the elimination period.

With this framework, our initial verification confidence will increase with time due to the aging and "enforced" inactivity of any possible covert missile systems.

2.2.2 The Details

The key verification elements are the following: production and test ban; continuous portal monitoring; restriction of INF missiles and launchers to designated facilities and areas; deployment areas; transit notifications; elimination procedures; baseline inspection; short-notice inspection of both active and de-activated missile support facilities and operating bases.

Production and Test Ban. Production of missiles, launchers, and stages was banned upon entry into force of the Treaty on June 1, 1988. Flight testing is also banned. The plant in Votkinsk that has assembled the SS-20 is being monitored by a continuous US on-site presence, which should ensure that SS-20s are no longer being produced there. Thus, the Soviets would be forced to turn to a covert assembly line if they sought to produce SS-20s. If they did so, they would need to "qualify" that covert line by testing the missiles or accept great risk through uncertainty in performance. However, such testing would be readily detectable by US NTM. The production ban, test ban, and continuous monitoring at Votkinsk act in concert to help deter cheating.

Continuous Portal Monitoring (including perimeter patrol). For up to 13 years the USA will have the right to station inspectors permanently outside any plant that assembles the SS-25 (or any missile that has a stage similar to a stage of the SS-20). Currently, the only such plant is in Votkinsk, which also was the assembly plant for the SS-20. The Soviet side has a reciprocal right at the Magna, Utah plant, which, at one time, produced the Pershing II missile.

Restriction of INF Missiles and Launchers to Designated Facilities or Areas. Missiles and launchers may be located only in designated facilities or areas, except when in announced transit. This partitioning of the Soviet Union and its basing countries into permitted and prohibited areas reduces the verification task to more manageable proportions. "*Counting*" would be accomplished chiefly by on-site measures at declared facilities and bases, small relative to the rest of the land mass of the USSR, East Germany and Czechoslovakia. In the remaining areas of these countries, the verification task is *detection* of any covert INF systems. Even one such missile or launcher detected in the prohibited area would be a violation.

Deployment Areas (DAs). During the reduction period, geographical restrictions will provide a verifiable basis for distinguishing deployed from non-deployed INF systems. Deployed intermediate range missiles (IRMs) are those located in specified areas where they may legally operate and exercise. Deployed short-range missiles (SRMs) were those located at their declared operating bases until they were "corralled" into special elimination facilities within 90 days after entry into force, i.e., by September 1, 1988.

Transit Notifications. In order to help resolve ambiguity regarding whether or not an INF missile or launcher outside a DA or missile support facility is in legal transit or is an illegal system, the Treaty requires the sides to provide information, after the fact, on the departures, arrivals and specified points in between for all such movement.

Elimination Procedures. For missiles, launchers, support equipment, and support structures listed in the Memorandum of Understanding (MOU), elimination is virtually synonymous with their physical destruction. The elimination protocol details procedures to achieve permanent destruction. In general, missiles may be destroyed by burning or demolition (or, in the case of up to 100 IRMs for a six-month period, launched to destruction); and launchers and launch-critical support equipment must be irreversibly altered so as to be incapable of transporting or launching IRMs.

Because of the small size and mobility of systems covered in the INF Treaty, elimination will be confirmed by on-site inspection during the actual elimination itself.

In most cases, it would be impractical or impossible to destroy missile support facilities, deployment areas, and operating bases; instead, elimination means that all INF activities there will cease. To help verify this, the Treaty provides for two types of interior OSI:

- *"Close-out" Inspection.* This is a one-time check of all facilities except missile production facilities to help ensure that all bases and support facilities have ceased any INF activity and that no missiles or launchers are present.
- *Short-notice Inspection at Formerly Declared Bases and Facilities.* All facilities and bases eliminated and subject to the one-time close-out are also subject to subsequent short-notice inspection to help deter the reactivation of former INF infrastructure to support a covert force. The treaty provides for a quota of such inspections for ten years after global elimination.

Baseline Inspection. A very detailed exchange of data, contained in the Treaty's MOU exchanged on signing (December 8, 1987) and initial data update on entry into force (June 1, 1988), establishes the baseline declarations against which subsequent compliance judgments will be made. The MOU is intended as a "snap shot" of the announced status of INF shortly before the Treaty was signed. In it, each side declared the number and location of each treaty-limited item, technical details about these items, and other pertinent data.

The purpose of the baseline inspections is to confirm the overt inventory at declared facilities, and thereby encourage truthful declarations. The baseline inspections would force the parties to hold any illegal missiles at covert locations, thus potentially exposing them to NTM detection.

Short-Notice Inspections of Active and Closed-out Missile Support Facilities and Operating Bases. All support facilities (except missile production) are subject to short-notice inspections. A yearly quota of up to 20 such inspections may be conducted during the 3-year elimination period, dropping to 15 per year, and then 10 per year during the two five-year periods following June 1991. These short-notice inspections will help deter the parties from using or re-activating the existing infrastructure to maintain and operate illegal INF systems.

2.2.3 Implementation

From the verification concept and elements I shall move to implementation via the key INF verification tasks and the specific measures for handling them. Some of the measures include: verifying the veracity of the initial declarations in the MOU and initial update; verifying compliance with numerical limits on missiles and launchers; verifying that all declared systems are eliminated as required; and verifying compliance with the ban on production and flight-testing (effective upon the Treaty's entry into force).

Baseline Verification. Starting 30 days after the Treaty's entry into force, the USA has conducted, over 60 days, inspections at about 125 sites in the USSR, GDR, and Czechoslovakia, using up to 200 inspectors in teams of 10 people each. The Soviet side has similar rights in the USA and five NATO basing countries. Subject to inspection are all INF missile support facilities, except missile production facilities and all missile operating bases. The former category includes missile and launcher repair facilities, missile and launcher storage facilities, test ranges, and elimination facilities.

We did not expect actually to find discrepancies as the result of our inspections. Rather, we view these inspections as a means of encouraging truthful declarations in the first place, i.e., as a deterrent to non-compliant behaviour.

We must assume that the other side *may* try to hide additional missiles covertly. Given the bi-lateral history, the relative small size and mobility of these systems, and past Soviet practices, we should not rule out this possibility. Indeed, the USA is devoting substantial monitoring resources to searching for any evidence of covertly-held INFs.

Verification of Numerical Limits on Missiles and Launchers. Verification of numerical limits is keyed to geographical restrictions, but these restrictions apply differently to the two major classes of INF systems - IRMs and SRMs. Any missile or launcher detected outside declared areas, and not in legal transit, is a violation. The transit notifications described earlier will help to distinguish legitimate transit from the operation of a covert INF system outside declared areas.

SRMs are defined as deployed if they are located at a missile operating base. All deployed short-range missiles and all launchers must be corralled in their separate elimination facilities within 90 days of entry into force. Non-deployed SRMs need not be moved to corrals for 12 months. Elimination facilities for short range missiles and launchers must be at least 1,000 km apart. Once there, they may not leave, and all must be destroyed within 18 months of entry into force.

The advantages of this scheme for controlling deployed SRMs are two-fold. First, since all SRMs are taken out of deployment status in 90 days or less, there will be no overt deployments to count. Second, since missiles and launchers are segregated in corrals 1,000 km apart, and subject to OSI, these systems cannot be easily redeployed without detection.

IRM reductions are more gradual, as dictated by military and political exigencies. IRMs must continue to exercise to remain a viable force until they are eliminated in three years. Deployment areas were devised to control and define deployment for verification purposes. Deployment areas (DAs) are designated areas within which IRMs and their launchers may operate and within which their operating bases are located. Every intermediate range missile or launcher within a DA is considered deployed, regardless of its operational readiness. The DAs are large enough to provide survivability for these deployed systems. Such protection for survivability in deployment areas for a period of three years is expressly permitted by the Treaty.

Short-notice inspection has a major role with regard to non-deployed limits. Non-deployed systems are confined chiefly to storage, repair, and elimination facilities. The key rules for the conduct of short-notice inspection are:

- Inspections may be made only within pre-specified boundaries, delineated in detailed site diagrams exchanged between the sides.
- The inspectors have the right to inspect anything big enough to be or contain a treaty-limited item.
- The inspectors can patrol the perimeter of the site and inspect outgoing vehicles.
- The burden of explanation is on the inspected party.

Equipment allowed for inspections includes measuring and weighing devices, cameras, and radiation detection devices.

Verification of the Production Ban. Beginning immediately upon entry into force, the sides are forbidden to produce INF missiles or stages, or to launch or flight-test the missiles.

The USA has long been aware of similarities between the Soviet two-stage intermediate-range SS-20 missile and the three-stage SS-25 intercontinental ballistic missile. Late in the negotiations, the Soviets revealed that the first stages for the two missiles, while not interchangeable, are very similar in external appearance. This posed a special problem for verifying a production ban. To ensure that the Soviets are not producing SS-20s in the guise of SS-25s, or other ballistic missiles not limited by this treaty, the USA negotiated the right to maintain a long-term continuous monitoring presence at the gate of any plant which assembles a ballistic missile using stages, any one of which is outwardly similar to a stage of a solid-propellant ballistic missile listed in Article III of the Treaty (e.g., the SS-25). This continuous monitoring right will last for 13 years if the Soviets continue to produce SS-25s, or any other missile with a stage outwardly similar to a stage of the SS-20. If the Soviets cease such production, the monitoring period would be shorter, but in no case less than three years after the Treaty enters into force.

The Soviet Union has assembled SS-25s and SS-20s at the same plant. That plant is at Votkinsk, approximately 1,000 km east of Moscow. We have, since July 1, 1988, implemented the right to continuous portal monitoring at Votkinsk. The key features are:

- Road and rail vehicles large enough to contain the SS-20 may exit only at one designated "portal", at which inspectors and various inspection devices are located.
- There may be only two other exits, each of which will be monitored by remote means, such as video cameras.
- The perimeter of the facility may be patrolled by inspectors to ensure that no incursions have occurred.
- Every vehicle exiting a portal may be weighed and measured to determine if it is capable of containing an SS-20. Vehicles incapable of containing such a missile will pass without further inspection.
- Any vehicle that is large and heavy enough to contain an SS-20, but is declared by the Soviets not to contain a missile, may be inspected. It is the responsibility of the Soviets to show that the vehicle does not contain a missile.
- Vehicles exiting the portal, which are declared to contain or could contain an SS-20 missile, are subject to the following procedures:
  - The USA may image the contents of each missile canister (e.g., by radiography).
  - The USA may select eight canisters each year for the Soviets to open for interior viewing. As with X-ray imaging, the viewing procedures will enable us to distinguish an SS-20 from an SS-25 by verifying that the second stage of the SS-25 is in fact different in dimensions from the second stage of the SS-20. The two viewing means will serve as check and calibration, one means against the other.

As discussed before, the flight-test ban is designed to strengthen the production ban by making it very difficult, if not impossible, to qualify a covert assembly line. Verification of the flight-test ban will rely on highly capable NTM systems.

Over time (e.g., 13 years), with the flight-test and training ban, and the limited shelf life of solid propellant missiles, it will be very difficult, if not infeasible, to maintain a militarily-useful covert force.

Verify Elimination of Missiles, Launchers and Support Equipment. The protocol on elimination contains detailed procedures to:

- Ensure complete and permanent destruction.
- Verify the number and type of items being eliminated.
- Verify the fact of destruction.

Specific procedures are prescribed for each type of item to be eliminated; thus, the elimination protocol serves as a destruction manual. Missiles, launchers and support equipment are destroyed at special elimination facilities, where inspectors may observe the whole destruction process from start to finish. Alternatively, a quota of 100 IRMs can be destroyed by launching in the first six months. Training equipment and support structures will be destroyed where they are located, and inspectors may observe the results of destruction after the fact.

The following are some of the specific means by which INF systems will be destroyed. The SS-20 will serve as an example. The sides must notify 30 days before they begin destruction of INF systems, providing information on the number and type, the elimination site, the location from which removed, and the point of entry for our inspectors. Each side is permitted 20 inspectors in the team that carries out the elimination inspections.

Prior to the missiles' arrival at the elimination facility, the inspected party will remove their nuclear warhead and guidance packages. However, the necessary "reentry vehicle" (i.e., heat shield for the warhead) will be delivered to the elimination facility. The SS-20 missile is always contained in a "canister" which provides the missile's environmental control. Just prior to a missile's destruction, the US inspector may direct the Soviets to open this canister so that he may visually inspect its contents to ensure that an SS-20 is actually inside.

The Soviets have the option of destroying the SS-20 at the facility by burning its fuel or demolishing the missile. Any fuel, nozzles, or motor case surviving this process must be burned, crushed, flattened, or destroyed by explosion. The RV must also be destroyed, by crushing or flattening.

The SS-20 transporter-erector-launcher will be stripped of its erection/launch device - basically the hydraulic erection arm and the missile carriage. All components of this device must be cut at locations that are not assembly joints. Mountings and leveling supports will also be removed from the transporter chassis and cut as above. Additionally, at least 0.78 meters must be cut off the chassis. As a result of these procedures, all launch-critical equipment is permanently destroyed and the structural integrity of the remaining transporter for erecting its missile is destroyed. (This altered transporter can be used for other purposes.) US inspectors may observe each step described. The vehicle used to transport resupply missiles and load them onto launchers will also be destroyed, as will simulator missiles and launchers used in training.

The Soviets may, for destruction purposes, launch up to 100 SS-20s in the first six months. The USA is allowed to inspect each missile prior to launch and the Soviets are prohibited from telemetering test information on these flights. However, because such launchings provide information on missile performance regardless of telemetry, the USA insisted on the six-month restriction and the quota. Our objective was to start the clock ticking towards INF missile obsolescence as soon as possible.

2.3 Status of Treaty Implementation

2.3.1 The On-Site Inspection Agency (OSIA)

Implementation of the complex on-site inspection regime for INF required the creation of a new entity known as the On-Site Inspection Agency (OSIA). This agency was established on January 15, 1988 as a field-operating agency of the Department of Defense.

The OSIA mission consists of inspection and escort responsibilities. This involves the management of US INF-related on-site inspection activities in the Soviet Union, Czechoslovakia and East Germany, and the coordination of activities associated with the Soviet Union's inspection of US INF facilities in the United States and Western Europe (Great Britain, Netherlands, Belgium, Italy and the Federal Republic of Germany). The OSIA is also responsible for continuous monitoring of the Soviet missile assembly facility at Votkinsk, and coordination of the Soviet Union's reciprocal monitoring of a US missile production facility at Magna, Utah.

The OSIA consists of personnel from all of the armed services and civilian experts and support personnel. It maintains close liaison with the interested government agencies, and draws its three deputy directors from the Arms Control and Disarmament Agency (principal deputy), the Department of State, and the Federal Bureau of Investigation.

2.3.2 The Inspection Process

Considering the complexity of the inspection regime and the historically adversarial bi-lateral relationship, the INF inspection process has been proceeding remarkably well. Technical problems and some misunderstandings have occurred and some technical details remain to be worked out. But, overall, the results so far are impressive.

The 60 days of baseline inspection were particularly hectic for the US inspectors who had to cover more than two inspection sites per day for a two months period. That phase was completed on August 1 with generally satisfactory results. Close-out inspections, short-notice quota inspections, and elimination inspections have followed. We are inspecting rail cars exiting the Votkinsk portal - to be sure that no SS-20s pass through - and are preparing to install there the Cargoscan X-ray imaging system - a commercially available technology. The USA has twice exercised its right to open SS-25 canisters.

The Soviet inspectors are installed at the Magna, Utah, portal and they have also conducted numerous baseline, close-out, short notice (quota) and elimination inspections on US and basing country territory.

2.3.3 The Special Verification Commission (SVC)

With such a complicated Treaty (about ten times the total length of any previous nuclear arms control treaty) and verification regime, it was clear to both sides that a forum would be needed to discuss treaty issues. The Treaty (Article XIII) creates a Special Verification Commission (SVC) as one means for resolving questions relating to compliance and for agreeing on any measures which may be necessary to improve the viability and effectiveness of the treaty. The SVC meets on request of either party, and may be at varying policy and technical levels. So far two sessions of the SVC have been completed (as of November 15, 1988). At SVC-II, discussions continued on a major Memorandum of Agreement with numerous detailed annexes. The purpose of these documents would be to lay out in detail procedures for implementing the rights embodied in the Treaty.

2.4 Verification Problems in Strategic Arms Reductions

While there is no doubt that much will be learned from the INF Treaty experience that would be applicable to a strategic arms treaty, the strategic regime is vastly more complicated and difficult. There are a number of reasons for this:

- First of all, there are many more types of weapons to deal with: ballistic missiles - fixed, land-mobile, and sea mobile (SLBM); heavy bombers - with and without air-launched cruise missiles (ALCM); and other cruise missiles (e.g., SLCM).
- Secondly, we are not banning a class of weapons, as in INF, but only reducing their number; there will be significant numbers of this weapon class remaining so that production and flight testing will continue. Also, the sides would be permitted to develop new types.
- Thirdly, there will be treaty constraints on "qualitative" factors such as throw-weight (TW) and numbers of reentry vehicles (RVs). This requires verifiable definitions of throw-weight preferably based on capability rather than demonstrations.
- Fourthly, these weapons are at the heart of each nation's strategic considerations. If a party does not live up to its obligations then the strategic deterrent and balance could be impaired. The INF systems did not pose such risks.

This leads to a number of difficulties not present in the INF "zero-zero" regime. For example, if mobile ICBMs are "permitted" in START, then a fixed number of systems must be verified, not just ensuring that none exist. "Counting" weapons systems is an inherently more difficult job than ensuring "zero". The absence of production and flight test bans will require much more extensive on-site inspections of all kinds including expanded perimeter-portal monitoring of production facilities.

The strategic weapons negotiations continue to focus on the limitation of the large, highly fractionated ICBMs (the "fast flyers") which are at the core of a de-stabilizing first-strike capability. Nonetheless, the "slow-flyers", such as heavy bombers with and without ALCMs, will also be limited by START and will be subject to verification provisions. Thus, we need viable, verifiable definitions of, say, what constitutes a heavy bomber. Effective definitions will often require meaningful surrogates of military capability; these surrogates should, preferably be measurable by NTM.

It might be helpful to wait for more INF Treaty implementation experience before completing a strategic forces limitation and reduction treaty. But political and popular pressures will probably not afford such a luxury. We are already deep into START negotiations and there is probably no turning back. And, indeed, the risks of enduring a longer period without strategic arms reductions may well out-weigh the potential benefits of logging more INF implementation experience.

2.5 Concluding Remarks

The fundamental objective of strategic arms reductions is to achieve higher levels of strategic stability at the lowest possible level of forces. The INF Treaty implementation process will be a unique test bed for the kinds of on-site inspections (OSI) and perimeter-portal monitoring (PPM) which will also form the basis for START. But, as indicated in the preceding section, the on-site inspection rights and the number of sites subject to PPM will expand greatly. A much larger OSIA organization may be required and the higher costs of this greatly expanded inspection regime will have to be addressed. The conflicting demands of verifiability vs. security and operational flexibility will be challenges to both parties to the future treaty.

The evolutionary goals of verification, beyond the immediate objectives of being able to assess compliance, include:

- Convincing the US public and the Congress that effective verification is being achieved.
- Building mutual confidence in both parties which could allay some of the fundamental distrust which 70 years of adversarial relationships has engendered.

It is difficult to conclude this discussion without some speculation on the potential social and political impact of these unprecedented exchanges of personnel. The opening up of sites and systems which the INF parties have for decades taken great pains to conceal from each other and the attendant exposure - albeit carefully circumscribed - cannot fail to have some lasting impact on the participants if not their publics at large.

These mutual exchanges which we are witnessing may be having effects which even short term changes in the political winds may not be able to hold back in the longer term. While carefully avoiding undue and premature euphoria, both sides need to observe carefully and seek out the important lessons to be learned in this unfolding process.

2.6 Notes

1 The views contained in this paper do not necessarily represent those of the US Arms Control and Disarmament Agency or any other US government agency

3. Experience with INF Treaty Verification and Prospects for the Future

Viktor Koltunov

3.1 Verification and Arms Control

The achievement of international agreements in the limitation and reduction of armaments is inseparably connected with the solution of the problem of verifying these agreements. Since we are concerned with agreements of vital interest to the security of the states it is extremely important to ensure that all participants comply strictly with these agreements.

With regard to the methods of solving the verification problem, one can discern two fundamental approaches.

The first puts the initial emphasis on the verification issue. Measures of verification are considered to be the essential part of the problem, whereas the elaboration of concrete measures of arms reduction and arms control are given lower priority. In other words, verification is seen as an end in itself, isolated from the measures of arms control and reduction.

The second approach - and in modern times this is the fundamental approach - puts the emphasis on arms control agreements; methods of verification are seen as a means of ensuring compliance with signed agreements. It could not, and should not play a self-sufficient role in this process. The volume, the design, and the measures of verification should be determined by the nature, the volume of the means of disarmament. The issues of verification, and those of reduction, arms control, and disarmament, should be analysed in parallel. If treated separately from the process of disarmament, verification becomes meaningless.

Long experience gained during negotiations on different aspects of arms control has shown, that the practical implementation of a balance between the measures of verification and arms control is not always simple, especially if this involves complex technical problems. However, if there is genuine interest on both sides of the negotiations to achieve agreements acceptable to both sides, then solutions can be easily found. This is borne out by the experience of many agreements, such as those relating to START negotiations, INF negotiations, CTB negotiations and others. And vice versa, the experience gained in negotiations shows that when there is no readiness to achieve an agreement, the problem of verification is often used as an excuse to prolong the negotiations and impede an agreement on arms control and arms reduction.

There is yet another very important principle which had proved to be correct in practice. The verification of disarmament is a problem which lies in the sphere of international relations. Consequently, it should be based on the fundamental principles of international law as stated in the United Nations Charter, such as sovereign equality and non-interference in the internal policies of the states. No system of verification can be considered legitimate unless based on these principles.

3.2 Verification of the INF Treaty

Let me now speak about the measures and methods of verification. In the 60s and 70s radical improvements were made in the elaboration and performance of various electronic, photographic, seismic and other means of verification. The national technical means thus acquired a fundamental importance from the viewpoint of ensuring compliance with agreements. One of the main methods of verification by national technical means are space satellites with different devices on board. We could say that the creation of these space satellites has made possible the conclusion of a number of agreements.

However, with the increasing complexity of the subject of negotiations, it became clear that national technical means alone are not sufficient. As we approach real disarmament and a real reduction of strategic nuclear weapons, the centre of gravity in the problems of verification shifts to on-site inspection, a different type of inspection. This was seen, in particular, in the experience gained from Soviet/American negotiations on the INF agreement. This agreement is unique in respect of the volume of the reduction of armaments. It provides for the complete abolition of two classes of nuclear weapons, with an elaborated mechanism for verification.

The system of verification provided in this Treaty also includes on-site inspection on a permanent and a challenge basis, the use of national technical means, exchange of data, information about the means envisaged in the agreement. To facilitate the realization of the objectives and the principles of the agreement, a special control commission, was created - a Special Verification Commission for the purpose of solving any questions relating to the fulfilment of the obligations entered into by both sides of the agreement and for coordinating measures. This commission was also charged with the task of facilitating the viability and efficiency of the Treaty.

The system of inspection under the INF agreement includes several types of on-site inspection:

- inspections dealing with verification of initial data (these inspections have already been carried out);
- inspections (to be carried out within a period of thirteen years) of all missile operating bases and missile support facilities;
- inspections on a *permanent basis* of plants and factories which were previously producing missiles (one plant or factory for each side);

- inspections dealing with the verification of the elimination of rockets and launching devices;
- inspections of the process of elimination of the missiles themselves and their delivery systems.

The inspections are carried out on the territories of the USA and USSR, as well as on the territories of the countries where the missiles to be eliminated are situated: GDR and Czechoslovakia; FRG, UK, Italy, Belgium and the Netherlands.

What does the experience of realizing the INF Treaty show? It shows that the means of verification provided in the Treaty give full assurance that all the statements about verification made in the Treaty are fulfilled by the participants of the agreement. Both sides have by now eliminated approximately 500 missiles and there was not a single complaint on either side, although some questions on technical details did arise. But those questions were easily discussed and solved on a cooperative basis.

3.3 Future Prospects of Verification

It seems to me that the statements, and the INF Treaty itself in the sense of verification measures foreseen by the Treaty, provide a good example and a good basis for the creation of a system and procedure of verification in future agreements. It should also be mentioned that they have found some reflection and development in the preparatory stages of the agreement on the 50% reduction of strategic nuclear weapons between the USSR and the United States. And some of them, according to the opinion on the Soviet side, could probably be used for verification of long range ballistic missiles.

As mentioned above, concrete forms of verification of disarmament measures are determined by the nature of the arms reduction measures and, of course, there is a great variety of methods, means and procedures of verification. The whole process of verification and its components is in constant flux, and this makes it virtually impossible to develop any ready-made recipes for future treaties.

However, on the basis of the experience gathered during many years of the functioning of the verification system, one can pick those components which have proved themselves, and utilize them in developing future understanding in the area of disarmament and arms reductions.

In our opinion the national technical means of verification will not loose their efficiency and importance in the future. And since the technical characteristics of space and other remote-sensing devices are increasing from year to year, we hope that the possibilities and the potential of the national technical means will grow further. Moreover, measures already brought into action, for example, in the framework of the INF agreement, which will probably be more widely practiced in the future, would raise the effectiveness of national technical means.

For example, one could think about an open exhibition of weaponry for a certain period of time at a certain location, of the means foreseen under the Treaty, the place and time of such an exhibition being chosen by the other side. The same purpose, e.g. the raising of the effectiveness of national technical means, is also served by the prohibition of jamming systems of recording of telemetric information and of the use of premeditated measures of concealment.

We think that the most effective method of verification for all sorts of agreements would be on-site inspections. We should note, however, the importance of ensuring an equitable approach by both participants for the agreement. For example, when we discuss the convention dealing with the prohibition of chemical weapons, the verification procedure should embrace all the objects in the states participating in the convention which are under suspicion, no matter to whom they belong and where they are situated. This approach is of course justified for any other agreement.

The experience of organizing on-site inspections on a regular basis points to the need of additional provisions for the improvement of such inspections, both from the standpoint of technical equipment and lowering the cost of these inspections.

I would also like to stress one special category of measures which help to support the whole process of the international stability of our societies. These, so-called confidence-building measures, aim at increasing confidence in the military field. They are not, strictly speaking, verification measures. They are based on the principles of good will and reciprocity. Yet, those measures enable the countries to acquire some understanding about different types of military activity of the parties participating in the process, for the purpose of creating trust and understanding about the plans of the other party. Some confidence-building measures, for example, information about military exercises, exchange of observers during exercises, have already been implemented. Other measures should be discussed and agreed upon during future negotiations on the limitation of conventional arms in Europe, from the Atlantic to the Urals.

Effective verification of the process of disarmament can be achieved by *special verification organizations* or by international organizations with verification functions, with the consent of the given organizations. In the first place we should bear in mind (and utilize) the tremendous experience acquired by IAEA during its existence.

The participants of the agreements can also organize periodic *conferences* for the examination of compliance with the agreements, and the whole procedure of verification and its efficiency. Such conferences could be very useful for international experience in elaborating the procedure of verification itself.

In summary, there are valid reasons to believe that we have at present quite a rich arsenal of different forms, methods, means and procedures of verification, some of which had been tested in practice and proved to be highly satisfactory.

4. Mechanisms for Raising and Resolving Compliance Issues

Gloria Duffy

4.1 Compliance Problems

Arms control compliance is the substance of arms control. Arms control agreements for the purpose of stating norms are important, but they are fairly meaningless if compliance is poor.

The compliance record is not as poor as either the USA or the Soviet Union have made it out to be throughout this last decade in their official statements about one another.[1] The overall pattern of USA and Soviet behaviour, when looked at as a whole, indicates that compliance has been quite good. But there have been some significant problems: the Krasnoyarsk radar, Soviet encryption practices, the US interpretation of the ABM Treaty, the yellow rain controversy. Even if many of these are not clear violations of arms control treaties, still, they are unresolved disputes; and they have been destructive and even contributed to the abandonment of treaties, as in the case of the USA abandoning the SALT I and SALT II treaties in 1981 and 1986.

We are now in a period of great enthusiasm for arms control, after the INF agreement, and so there has been a tendency to forget about these very recent compliance problems. But problems of these types will reappear, for two reasons. First, because it is simply impossible to negotiate and conclude an arms control agreement that avoids all disagreements on implementation and interpretation. This is not possible in human endeavour. In particular, compliance disputes are bound to reappear during the next difficult period of USA-USSR political relations, because when political relations are poor, divergent interpretations of the terms of agreement are more common, and the resolution of disputes is more difficult.

Second, because the mutual security regime is expanding, year to year. Even with the setbacks experienced during the period of USA-USSR stalemate, from the mid-seventies to the mid-eighties, the overall historical trend is one of expansion in mutual security arrangements. In addition to the twenty or so major East-West security agreements now in force, there is the prospect for perhaps five more agreements over the next decade: on East-West conventional arms control; a chemical weapons ban or further limitations; testing limits of some sort; a 50% or more reduction in strategic forces; and clarification or renegotiation of the ABM Treaty to deal with threats like SDI and other defence programmes.

In addition, there are other agreements, such as an ASAT ban, or for US-Soviet regional disengagement, that are less likely than these five, but still possible. Thus, between the existing agreements and the potential new ones, we have an expanding web of treaty provisions and obligations that will require implementation, clarification, and adjudication where disagreement will inevitably occur. Measures to deal with compliance issues will be, and should be, consuming ever more of our attention.

The compliance mechanisms that exist today are a hodge-podge of different measures for different treaties. They range from agreements that have no formal compliance mechanisms or dispute resolution mechanisms at all, as in the case of the Limited Test Ban Treaty, to very intensive measures, such as the use of the Nuclear Risk Reduction Centres for data exchange and the Special Verification Commission for dispute resolution, the provisions designed for the INF Treaty. The provisions for existing agreements range from vague requirements for consultation, as in the cases of the Seabed and Outer Space Treaties, to bilateral consultative mechanisms, some of which, as in the case of the Joint Consultative Commission for the Threshold Test Ban Treaty and the Peaceful Nuclear Explosions Treaty, have never come into force. They include multilateral mechanisms, as in the case of the NPT, or the Stockholm and Helsinki agreements which include international mechanisms as well. A number of treaties call for resort to the UN Security Council, or involvement by the Secretary-General, as in the case of the Biological and Toxin Weapons Convention. Other agreements, the Antarctic agreement, for instance, call for disputes to be referred to the International Court of Justice for resolution.

Thus, the current provisions for compliance and dispute resolution cover a wide range, from none at all to the involvement of international institutions. Today there is clearly a sense that something more than this scattered attention to compliance is needed. Too many agreements have no effective dispute resolution mechanisms. Such is the case of the Limited Test Ban Treaty and the Threshold Test Ban Treaty, which have left compliance issues to ad hoc arrangements between the parties, unable to resolve satisfactorily the recent disputes. And those compliance mechanisms which do exist can too easily be made useless, or be paralysed, by Soviet-American political tension, as occured from 1980 until very recently. Yet, it is precisely when tension is greatest that more disputes will tend to arise and mechanisms must function most effectively.

4.2 Possible New Compliance Mechanisms

What is this "something more" that we need? Many proposals have been made, most along the lines of further internationalizing the process. There is the Soviet proposal for a UN Verification Agency, and Soviet proposals for involving the Secretary-General in investigating allegations of cheating and resolving disputes. There is the French proposal for an international satellite verification agency. Most of these proposals concentrate on improving and internationalizing verification procedures and technologies and not on encouraging compliance or improving the dispute resolution procedures.

In my view, the present period, when East-West relations are not in a disputatious state, is a very good time to give thought to specific measures to foster compliance and to improve the mechanisms for resolving compliance disputes. In designing new mechanisms to improve the conditions which foster compliance, we have to deal with a number of questions. First of all, change or no change? Shall we change the current mechanisms at all, or keep them, assuming that as the political climate improves, the incentives to resolve disputes and the practices themselves, will also improve? Look at the positive trends in the USA and the Soviet Union: as political relations have become better, the Soviet Union has moved to deal with the Krasnoyarsk problem, and so forth.

The second question is the scope of any changes that might be made. What scope should new mechanisms have? Should they be simply improvements of the existing bilateral mechanisms, the creation and strengthening of multilateral mechanisms, or the creation of international mechanisms?

Third, what functions should the new mechanisms have? There are many functions that are related to treaty compliance, including implementation and data exchange; data collection, including on-site inspection; analysis of agreements, including treaty documents themselves and the negotiating history; consultation between the parties to agreements; arbitration and dispute resolution; and finally, enforcement or sanctions.

Fourth, should we strive for unity or diversity? Should there be one institution to deal with all agreements, to resolve compliance disputes, or should there be a different institution for every agreement, the mechanism which has been followed in general until now? Should the mechanisms operate publicly or privately? Should their deliberations and results be public or private? Should the decisions of compliance adjudicating bodies be binding or non-binding? Should they be consultative mechanisms or is there a way to structure them so that the results of their deliberations are more binding?

It is important to resist the temptation to change what we have simply for the sake of change. So in considering how to improve the present compliance mechanisms, it is important to think about the factors that encourage compliance, and then address those needs specifically by changing the system.

In my view, the factors which encourage compliance are:

1. a common understanding of treaty provisions;
2. treaty provisions which continue to be updated to the present context, and thus continue to be relevant to current behaviours and technologies;
3. a strong internal constituency for compliance, inside the countries party to the treaties;
4. a fear of sanctions if violations are discovered - a fear of discovery, response, perhaps public embarassment;
5. a mutual interest in making treaties work and a perception that one's own security will deteriorate if the treaty lapses;

6. a continued development and expansion of the mutual security regime, which creates incentives for compliance;
7. a higher authority to adjudicate disputes, once bilateral or multilateral efforts have ended in stalemate.

The absence of these conditions has contributed to the recent problems, such as the Krasnoyarsk issue, the US reinterpretation of the ABM Treaty, and the yellow rain controversy. Therefore, any new mechanisms should seek to improve these conditions. It is not possible to affect all of them by restructuring the process for handling compliance disputes. For instance, the continued development and expansion of the mutual security regime is outside the scope of institutional changes. But the rest of the factors can be manipulated.

Realistically, in changing the system that exists today, it is also important to balance the tension between the needs that countries have for autonomy and their desire for improved mechanisms for resolving disputes. The Soviet Union, for instance, has been in the forefront of proposing some new international mechanisms to deal with compliance disputes. In interviews that I did in the Soviet Union recently, I asked a number of people in the Foreign Ministry and in Institutes whether, for instance, the Soviet Union would have submitted the Krasnoyarsk issue for decision by some international agency with multilateral participation, and whether it would have taken to heart any decision rendered by such a body on the legality of the Krasnoyarsk radar. The answer in general was, "Well, it cannot be a binding judgement, but we would certainly have taken into account the opinion expressed by an international agency." So, both in the USA and in the Soviet Union there are barriers to relinquishing autonomy.

The most important goal of any new mechanisms to improve compliance is to de-politicize and further institutionalize the mechanisms for raising and resolving compliance disputes, in order to reduce paralysis in times of political tension. The only chance of being successful in doing this, is to build on the existing mechanisms, and to convince the parties that their interests will be protected by emphasizing expertise, neutrality, and arbitration in any new institutions that are created.

4.3 A Supranational Arbitration Agency

I would like to raise here a proposal that is being now discussed in Los Angeles by a delegation of the US Lawyers Alliance for Nuclear Arms Control, and a Soviet group of lawyers and jurists. It is a proposal originated by a US scholar, Phillip Trimble of the UCLA Law School, for an international institute for arms control dispute settlement.[2] He calls for nonbinding arbitration of all arms control disputes by a new institution which would absorb the Standing Consultative Commission, the Special Verification Commission, and all of the other current dispute resolution mechanisms. It would involve representation by official personnel of the countries involved - cabinet officers, and general staff members.

I would like to make a similar but somewhat different proposal. Because the need for autonomy is still high in resolving disputes, I would suggest that we maintain all the present mechanisms for compliance dispute resolution, adding compliance mechanisms to those agreements currently lacking them (the Limited Test Ban Treaty, the Threshold Test Ban Treaty). But, I would propose adding a supranational agency as a last resort to be used especially in times of political polarization. The reasons for suggesting such an "agency of last resort" to arbitrate and adjudicate compliance disputes rather than subsuming all existing mechanisms under a new agency are partly practical. As the arms control regime expands, it is simply not going to be possible to subsume all of the activities of all of the compliance mechanisms under one organization. But also, as the American scholar Jim Schear has written, "The strongest incentives to resolve disputes are by those directly involved". In other words, the countries most directly involved in a dispute have the most fear from the lapsing of the agreement, and the possible retaliation from the other side if a compliance dispute is not resolved. Due to this factor we can expect the success of existing compliance mechanisms most of the time.

An important function of the new international arms control compliance agency would be to dramatize the results of questionable behaviour, or non-compliant behaviour, for the security of the offending party. It would be important for such an agency to be able to release reports on the likely consequences of questionable behaviour, and responses to the questionable behaviour, that would decrease the security of the offender.

The functions of such an agency could include treaty interpretation and arbitration, the investigation of allegations of non-compliance, the development of expertise in dispute resolution, and dramatization through reports of the results of non-compliant behaviour. Such an agency could be planned by an international conference, and modelled on a panel of "wise men", or an international commission. Countries would subscribe to membership in the body, they would appoint representatives who would be legal scholars, former negotiators, diplomats with ambassadorial rank. I envisage this as a separate agency, not as part of the United Nations.

Such an agency would have the freedom to undertake examinations and investigations of its own, to analyse impending threats to treaty compliance, as well as to respond to requests for arbitration from the parties. It would develop expertise, perhaps through a training institute, in arbitration and dispute resolution, and eventually could funnel experts to the other subsidiary dispute resolution bodies attached to particular treaties. The proposed agency could have intelligence assets; access to satellites and other assets; it could do fact-finding, where relevant to specific investigations. It could be funded by assessment of the member states; it could determine its own voting requirements; it could release reports on the consequences of non-compliance if a dispute were not resolved, to highlight the security consequences of non-resolution of disputes. It would maintain confidentiality as long as a dispute were under discussion and being handled; within some limits it could be able to disclose some information, if the process of dispute resolution were to fail.

How would it work in a specific case? Let us take the case of the controversy over the high level of Soviet encryption, and compliance with the SALT II Treaty. As is known, this issue ended in stalemate at the Standing Consultative Commission. Now suppose that a suprabody for arbitration and dispute resolution existed. The USA could bring the issue of Soviet encryption to the new agency. The agency's personnel could examine the treaty provisions, and consider the positions of both countries. Perhaps the agency would recommend that the USA be more specific about its needs for data to verify compliance with the SALT II Treaty; perhaps it would recommend that the Soviet Union justify its high level of encryption.

If there were no change in the position of either side, which is where the issue has been languishing for several years now, the agency could report on the likely consequences for the security of both sides if the issue were not resolved. For instance, in the case of encryption, it could have identified the possible US abrogation of the SALT II Treaty with the attendant cost to the security of both the USA and the Soviet Union.

4.4 Notes and References

1 For a detailed analysis, see: Gloria Duffy (Project Director), Compliance and the Future of Arms Control, Stanford CA: Stanford University, 1988

2 Phillip R. Trimble, Beyond Verification: The Next Step in Arms Control, Harvard Law Review, vol. 102, pp. 885-912, Febr. 1989

Part II

Verification of Mobile Missiles, Nuclear Warheads and Fissile Material

5. Verification of Sea-Launched Cruise Missiles

Valerie Thomas

5.1 Introduction

It is widely recognized that sea-launched cruise missiles (SLCMs) are a difficult verification problem, and I am not going to present a "technical solution" to it. There is a variety of approaches which could be adopted for SLCM verification; those with the greatest advantages in terms of comprehensive monitoring also tend to have the greatest disadvantages in terms of intrusiveness and disruption of operations. Evaluation of these advantages and disadvantages tends to be idiosyncratic.

In this paper I shall analyze what the SLCM verification problems actually are, and the extent to which different verification approaches can eliminate or at least control these problems.[1] To this end, I will first discuss some of the general problems of SLCM arms control, and then the specific verification problems. Finally, I will discuss three general approaches to SLCM verification: maximal inspection, no inspection, and intermediate inspection.[2]

5.2 SLCMs and Arms Control

The United States has procured about 350 long-range nuclear Tomahawk SLCMs, and is planning a total of 758. The USA has also procured a much larger number of long-range non-nuclear Tomahawk SLCMs - about 1,100, with plans for a total of about 4,000.[3] Figure 5.1 shows the internal differences between the nuclear version of the Tomahawk and the non-nuclear versions: anti-ship, land attack, and land attack with multiple submunitions.

The Soviets are just beginning to deploy a long-range nuclear SLCM, the SS-N-21 reports indicate that the number is of the order of 100.[5] There has been no report of a non-nuclear version. There is also a supersonic Soviet long-range nuclear SLCM being developed, the SS-N-24. It will be large - about 13 metres long and 1 metre in diameter, in contrast to the SS-N-21 and the US Tomahawk, which are about 7 metres long and half a metre in diameter.[6] Therefore the SS-N-24 cannot be launched from torpedo tubes, but will require dedicated launchers, and is thus unlikely to be a difficult verification problem.

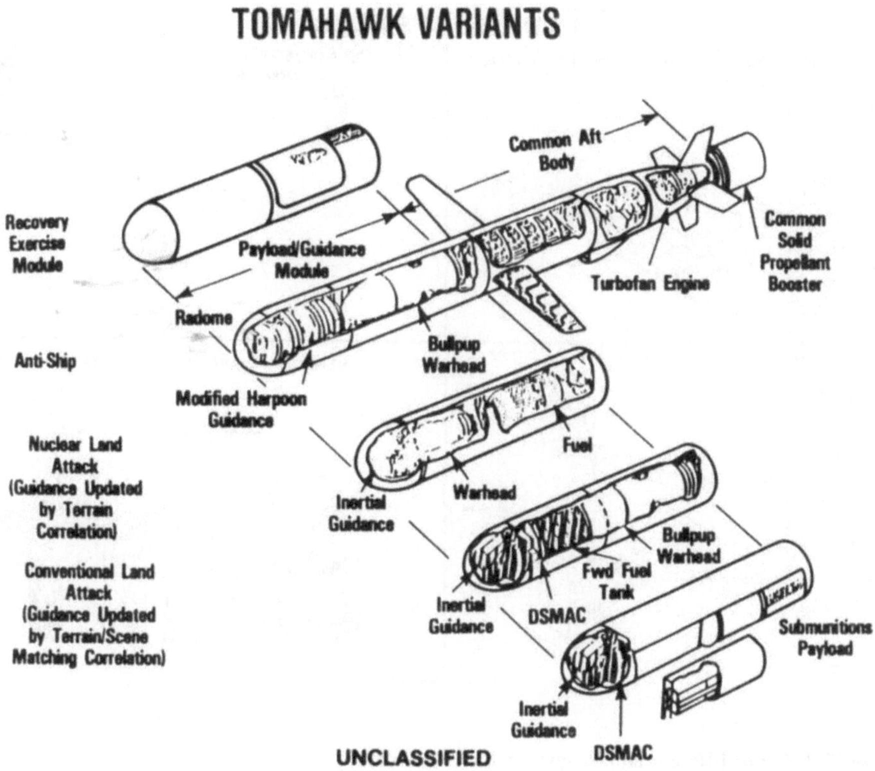

Fig. 5.1. Nuclear and non-nuclear variants of the US Tomahawk cruise missile [4]

Sea-launched cruise missiles remain an area of disagreement at the START negotiations. This can be attributed to a number of factors in addition to verification. First, it is the US view that cruise missiles should be treated differently from ballistic missile warheads because they are slower and do not pose a disarming threat, but the Soviets do not agree with this assessment.[7]

Second, the USA is ahead of the Soviets on SLCMs; the US has deployed more SLCMs on more types of vessels than the Soviets. A third and related point is that on-site inspection of SLCM limits is likely to be more disruptive of US Navy operations than of Soviet Navy operations. While the Soviets have reportedly deployed SLCMs only on dedicated cruise missile submarines, and have not developed a non-nuclear version, the USA will disperse SLCMs throughout the fleet, on about 100 ships and 100 submarines.[8] Over 80% of US Tomahawk SLCMs will be non-nuclear creating a need to distinguish the nuclear from non-nuclear versions. Also, the US

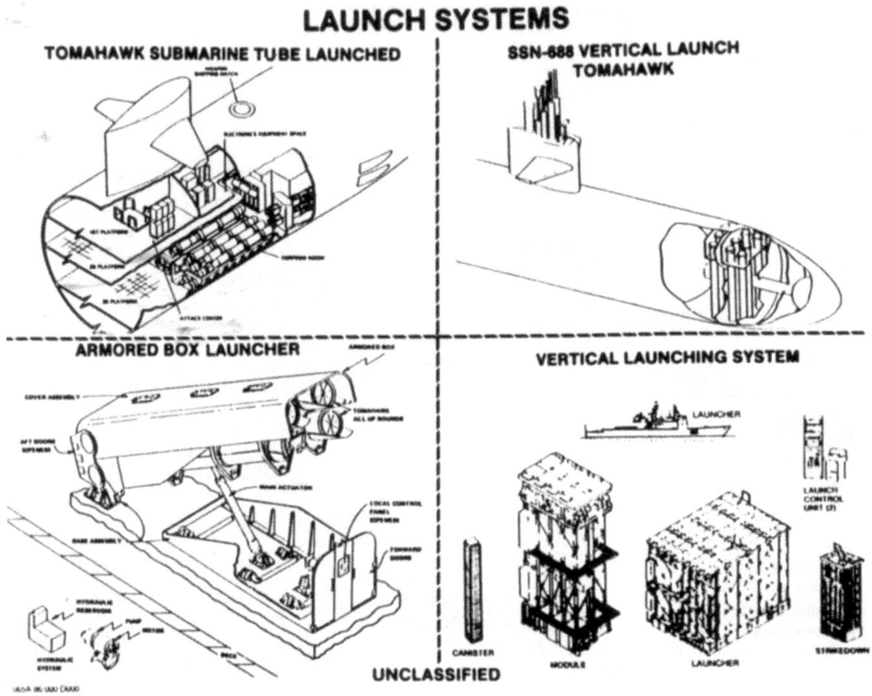

Fig. 5.2. Four different launch systems for US SLCMs [9]

SLCMs are stored in launchers which also store other (non-cruise, non-nuclear) naval weapons. Figure 5.2 shows the four types of launchers used for US Tomahawk SLCMs. The Vertical Launch System (VLS) also holds the "Standard" missile, and will hold a new, non-nuclear version of the ASROC anti-submarine missile. Torpedo rooms also hold torpedos, the Harpoon short-range cruise missile, and decoys. The submarine Capsule Launch System (CLS) and the Armored Box Launcher (ABL) are for long-range SLCMs only. To date, the Soviet SS-N-21 has only been reported on dedicated cruise missile submarines, but in the future it may be more widely deployed for torpedo tube launch. Soviet torpedo launch SLCMs give rise to the same verification problems as US torpedo launch SLCMs.

The United States has a policy of "neither confirming nor denying" the presence of nuclear weapons on its ships. This policy would have to be changed if the USA were required to reveal the location of its nuclear SLCMs.[10]

5.3 SLCM Verification Problems

1. The number of deployed US nuclear SLCMs cannot be counted by satellite, because some of the launchers are not dedicated to SLCMs, thus the number of launchers does not indicate the number of SLCMs. And since over 80% of US SLCMs will be non-nuclear, the correlation between SLCM-capable launchers and US nuclear SLCMs is even more tenuous.
2. Cruise missiles are easy to store, and easy to deploy, in comparison with other long-range nuclear weapons. Since they can be launched from all-purpose launchers, which are very unlikely to be eliminated by a treaty, stored SLCMs could become operational simply by being transferred to a ship. Therefore, absence of deployed SLCMs on ships and submarines ensures only a small time lag from possible treaty breakout to full scale deployment, if stored SLCMs are not eliminated. And, of course, it could be very difficult to identify any clandestine storage sites.
3. Other types of cruise missiles might be used as long-range nuclear SLCMs. Possibilities include installation of a nuclear warhead into a non-nuclear long-range SLCM, upgrading of the range of a short-range SLCM, or launching ALCMs (Air-Launched Cruise Missiles) from SLCM launchers. All of these possibilities are subject to certain constraints, however. The Soviets do not have a long-range non-nuclear SLCM, so at present the non-nuclear-to-nuclear conversion scenario does not apply to the Soviet arsenal. For US SLCMs, it would be possible to install a nuclear warhead into a non-nuclear SLCM, but this is reported to be a complex and time-consuming procedure.[11] As for upgrading of short-range SLCMs, the US short-range SLCM, the Harpoon (range: 56 km), is too small (4.5 m by 0.3 m) to hold enough fuel to become a long-range (600 km) SLCM.[12] But the Soviets do have a variety of short-range SLCMs, all of which are reported to be larger than the SS-N-21. The SS-N-12 and SS-N-19 are the longest ranged of the short-ranged SLCMs, with ranges reported to be 550 km.[13] It is conceivable that through use of a more efficient propulsion system and a light warhead, the range of these missile could be upgraded beyond the 600 km limit which defines a SLCM as long range. However, an upgrade of this sort would probably require testing, which might be monitored by national technical means.[14] As for the possibility of using ALCMs as SLCMs, this is probably not a serious concern for US ALCMs, since they have a maximum diameter of 69.3 cm, compared to 53 cm for the US SLCM.[15] Thus the US ALCM is too large to be launched from the standard 54 cm torpedo tube. Similarly, the VLS canisters can only accommodate missiles which can fit within a 56 by 56 cm^2 cross section.[16] The similarity of Soviet SLCMs and ALCMs has not been reported.

5.4 Verification Approaches

5.4.1 Maximal Inspection

Since SLCMs cannot be counted by satellite observation, and on-site inspection is now becoming almost routine, it is at least plausible that the best approach to verification is the most comprehensive one: to use on-site inspection to try to keep track of all the SLCMs, by monitoring production, maintenance, storage, deployment and destruction, and with challenge inspections of sites on land. Such an approach could be implemented as follows:

First, production sites would be continuously monitored, as would storage and maintenance facilities. At US facilities at least, non-nuclear SLCMs would need to be distinguished from nuclear. Inspection might be done at the factory gate, after the missile was in its canister. It seems unlikely that a nuclear warhead could be shielded within its canister, though this depends on the design of the missile and the nuclear warhead. In principle, there are several ways that the presence of a nuclear warhead, or its absence, might be detected. The warhead might produce detectable neutrons or gamma-rays, in which case passive detection measures would suffice. Possible active detection processes include backscattering of gamma-rays, gamma-ray or neutron transmission measurements, and induced fission.[17] Each SLCM could be labelled with an unreproducible tag or label, so that on subsequent inspections, each SLCM could be identified as one of the known number. If both nuclear and non-nuclear SLCMs were deployed, each missile canister could be sealed (with a tamper-revealing seal) to ensure, on subsequent inspections, that a nuclear SLCM had not been put into a previously non-nuclear canister.

The second aspect of the verification approach is deployment monitoring. It is probably unreasonable to expect full-scale searches of ships and submarines, or that naval vessels would be subject to inspections at any time and at any place. But perhaps ships at port might occasionally be inspected, with inspections limited to the area of the SLCM launchers. Alternatively, the Soviets have proposed permanent inspection teams at naval ports which would inspect and count each missile before loading onto a ship or submarine.[18]

The third component of this verification approach is challenge inspections of sites on land. Although challenge inspections are unlikely to chance upon a treaty violation, they do have the potential to reveal a suspected violation, or to clear up a misunderstanding at a site where questionable activities had been identified from intelligence sources.

There are some operational difficulties involved in the deployment monitoring (see Fig. 5.2); some SLCM launchers would be easier to monitor than others. The Armored Box Launcher (ABL) sits above deck. If the doors of the ABL were opened, inspectors could check tags on the missile canisters. Nuclear versions could be distinguished by a tag and perhaps a seal, or by fissile material monitoring.

The Vertical Launching System (VLS) presents more of a monitoring problem, since it also stores other types of weapons. But if the hatches were opened, SLCMs might be distinguished from other VLS weapons, since SLCMs are larger and heavier.[19] It is not clear whether it would be necessary to open the missile canister to distinguish SLCMs from other VLS missiles.

The submarine Capsule Launch System (CLS) could be inspected fairly easily. It stores only Tomahawk missiles, and cannot be reloaded from the inside of the submarine. The CLS hatches can be loaded opened to allow inspection from the exterior of the submarine.

SLCMs in torpedo rooms - to be launched from torpedo tubes - would be the most difficult to monitor. Direct inspection would require access to the interior of the attack submarine - which the US Navy considers highly sensitive. As an alternative to direct inspection, all weapons in the torpedo room could be brought out for inspection. Some assurance that all weapons had been brought out could be provided by requiring that a specified number - perhaps two dozen - of torpedo-sized weapons or other objects be brought out of the submarine for display. Since space is at a premium on submarines, keeping extra SLCMs on the submarine would be inconvenient, to say the least.

If SLCMs are to be inspected while in their launchers, any seal must be designed accordingly. A seal would have to be on the end of the missile canister in order to be inspected - though it would be necessary to seal the entire missile. So unless the canister can only be opened on the top, the seal would need to indicate a breach of any part of the canister. This might be done by a fibre optic or electronic seal.

As an alternative, or supplement to seals, it might be possible to detect the fissile material in the warhead - or confirm its absence - by measuring from the top through opened missile launchers. However, this would be more difficult than the fissile material detection at the production site, described above, since only one end of the missile is accessible; transverse scattering or transmission measurements would not be possible.

The difficulties of launcher inspection could be avoided by inspecting weapons at ports before they are loaded onto ships and submarines. This procedure has the advantages of avoiding direct inspection of ships and submarines, torpedo rooms in particular, and greater ease of inspection. The disadvantages are that every loading of SLCM-capable vessels would be monitored, whereas launcher inspections could be very infrequent.

This verification regime (inspection of SLCM launchers at port, monitoring of production, maintenance and storage sites, and challenge inspections on land) is the most intrusive regime that can be considered realistic. But even with so much inspection, there are possibilities for undetected violations or for sudden deployment after breakout from the treaty: 1. secret SLCM production cannot be ruled out because SLCM production facilities may not be particularly identifiable; 2. short-range SLCMs could be upgraded to long-range; 3. all previously manufactured SLCMs might not be declared; 4. even if non-nuclear SLCMs were sealed to prevent conversion to nuclear, the conversion could be done later if the treaty were broken; 5. ALCMs might be capable of being launched from SLCM launchers, in which case ALCMs could be diverted from the ALCM stockpile for use as SLCMs. The agreement would be strengthened by imposing comparable limits on ALCMs.

This verification regime also raises security and operational problems. The number of US sites to be inspected is comparable to the number of missiles. For example, if 400 nuclear SLCMs were deployed, 200 US vessels would be subject to inspection, plus all naval bases which handle SLCMs (of the order of 10), as well as any challenge inspection sites. With the exception of the production facilities, all of these sites would be related to other weapon systems and activities in addition to the SLCM. Some potential challenge inspection sites may be particularly problematic, such as production facilities for the stealth cruise missile, which is under development.[20] This technology is highly classified by the United States, yet the facilities which produce any type of cruise missiles are the most logical choices for challenge inspections. In short, this approach involves a lot of inspections. The USA has been unwilling to accept such a highly intrusive regime. Is there an acceptable alternative?

5.4.2 Minimal Inspection

The USA has proposed another approach: a "declaration of intent" on SLCM deployment.[21] The Soviets could declare how many SLCMs they will deploy, and the USA could do the same, and that would be that. No verification, no problems. This may not be as unverifiable as it sounds. After all, the Soviet Union does know how many nuclear SLCMs the USA has, from reading published US government documents. And the USA should have at least a fairly good idea of how many long-range nuclear SLCMs the Soviets have, since they are reportedly as yet deployed on only a small number of dedicated cruise missile submarines.

5.4.3 Intermediate Inspection

One can also search for a middle way: enough verification to be termed "effective", but not so much to be considered too intrusive. The goal would be to verify overt compliance - checking the number deployed and the number destroyed - and to rely on intelligence and national technical means to deter unlikely and complex scenarios, such as warhead switching or secret missile factories.

These criteria could be satisfied by monitoring declared production and maintenance facilities only. An initial data base of the number of SLCMs deployed could be exchanged. Any SLCMs to be destroyed would be destroyed in the presence of inspectors. For any remaining missiles, inspections would be at the production site for new missiles, or at the maintenance site for old missiles. At the first inspection, each missile could be tagged for future identification, and possibly sealed. On subsequent inspections - the end of every maintenance cycle - SLCMs would be inspected to check that only the tagged missiles passed through, and that the type of warhead, either nuclear or non-nuclear, was as designated by the tag. In the USA, SLCMs are brought back for maintenance once every three years.[22] On this schedule, after three years all SLCMs would be inspected and tagged, and reinspected at three year intervals.

There are obvious shortcomings with this method. There is no guarantee that all SLCMs will be brought to the designated maintenance site. But if the treaty were to be violated in secret, two separate sets of SLCMs would need to be maintained: the tagged and the untagged. None of the untagged SLCMs could be allowed to get mixed up with the tagged and sent to the wrong maintenance site. So quite a few people would have to know of the scheme - increasing the difficulty of pulling it off - or decreasing the likelihood that violation would be attempted. However, if the extra cruise missiles were kept stockpiled, rather than deployed, the difficulty of maintaining secret missiles would be minimized.

This intermediate verification regime would be strengthened by the right to challenge inspections anywhere on land that is suspected of housing extra SLCMs. For without challenge inspections, it could be difficult to determine whether suspicious activities indicate a violation or only a misinterpretation.

But meaningful challenge inspections would open up to possible inspection all naval bases and weapon depots, all cruise missile related production facilities, and other military installations. If such inspections were allowed, this regime might not seem significantly less intrusive than the maximal verification approach, yet it would probably be considered less reliable.

5.5 Conclusion

No matter what the START negotiators do about SLCMs, it is likely to be controversial. If they take the maximum verification approach, it may be seen as excessively intrusive, yet there remain scenarios for clandestine violation. Although these scenarios are of at best questionable plausibility, they are typical of verification concerns that have been raised in the United States for other arms control treaties.

If they choose the intermediate approach, there may be charges both that verification is not adequate and that the inspections are too intrusive. Indeed, this verification regime is not designed to prove treaty compliance, but rather to provide enough of a monitoring framework that it becomes implausible that serious violations would not be discovered by national intelligence. On this basis, it may be possible to form a consensus that this approach does provide adequate verification and that the inspections are not too intrusive.

If the negotiators choose the declaration of intent approach - essentially no verification - the issue of intrusiveness would not arise. Paradoxically, by deemphasizing the importance of SLCMs and their verification, this approach may minimize the political difficulties that a SLCM limitation would face.

5.6 Notes and References

1 A detailed study of SLCM verification will be published in: Frank von Hippel and Roald Sagdeev (eds.), Verification of Nuclear Warhead Reductions and Space Reactor Limitations, to be published by Gordon and Breach

2 See also: Potential Verification Provisions for Long-Range Nuclear-Armed Sea Launched Cruise Missiles, Workshop Report, Center for International Security and Arms Control, Stanford University, July 1988; and: Kosta Tsipis, Cruise Missiles Should Not Stop START, Bulletin of the Atomic Scientists, November 1988

3 Information provided by Cruise Missiles Project, U.S. Navy, reflecting FY 1988 estimates, March 11, 1988

4 Cruise Missiles Project, U.S. Navy; reprinted by permission

5 Jane's Defence Weekly, January 16, 1988

6 Soviet Military Power 1986, Washington: U.S. Government Printing Office, 1986, p. 34

7 Arms Control Reporter, pp. 611.B.488, 611.D.77, 1988

8 Statement of Adm. William C. Bowes, U.S. Navy, Director, Cruise Missile Project, before the Defense Subcommittee of the House Appropriations Committee on Tomahawk Weapon System, April 21, 1988

9 Cruise Missiles Project, U.S. Navy; reprinted by permission

10 Arms Control Reporter, p. 611.B.470, 1988

11 Testimony of Rear Adm. Stephen J. Hostettler, Director, Joint Cruise Missile Project, Senate Armed Services Committee, DoD Authorization for FY 1986, Part 7, March 8, 1985, p. 3875

12 T. B. Cochran, W. M. Arkin, and M. M. Hoenig, Nuclear Weapons Databook, vol. I, U.S. Nuclear Forces and Capabilities, Cambridge MA: Ballinger, 1984, p. 188. The range of the (non-nuclear) anti-ship Tomahawk (TASM) could potentially be upgraded, but this possibility would be taken care of in a verification program designed to distinguish nuclear from non-nuclear Tomahawks

13 Bernard Blake (ed.), Jane's Weapon Systems, 1987-88, New York: Jane's, 1987, pp. 484, 485

14 For a related discussion, see: The INF Treaty, Report of the Committee on Foreign Relations, U.S. Senate, April 14, 1988, pp. 60, 61

15 ALCM diameter from: Cochran et al. (note 12), p. 174. SLCM diameter from: McDonnell Douglas brochure (21 inches)

16 Vertical Launching System, Naval Sea Systems Command brochure.

17 See also: S. Fetter, O. F. Prilutskii, and S. N. Rodionov, Passive Detection of Nuclear Warheads, this volume, Ch. 6

18 Arms Control Reporter, p. 611.B.492, 1988
19 An Alternative to VLS UnRep, Proceedings of the U.S. Naval Institute, p. 129, December 1987; The Shield and the Sword, Sea Power, p. 25, February 1988
20 Aviation Week and Space Technology, p. 25, September 26, 1988
21 Arms Control Reporter, pp. 611.B.470, 611.B.493, 1988
22 Hostettler (note 11)

6. Passive Detection of Nuclear Warheads

Steve Fetter, Oleg F. Prilutskii, and Stanislav N. Rodionov

6.1 Introduction

Devices that could detect the presence of nuclear weapons could be useful in verifying compliance with various arms control agreements. Examples include monitoring a ban on nuclear weapons on surface ships, verifying compliance with the current ban on placing nuclear weapons in space, verifying limits on the number of nuclear warheads on missiles, or verifying limits on the nuclear versions of dual-capable weapons.

To the best of our knowledge, all nuclear weapons contain at least several kilogrammes of fissile material. Fissile material is the fuel for fission explosives such as those that destroyed Hiroshima and Nagasaki: it is also used in the triggers of modern thermonuclear weapons. The two fissile materials used in US warheads are highly enriched uranium (HEU) and weapons-grade plutonium (WGPu). The composition of these materials is summarized in Table 6.1.[1]

Fissile materials are radioactive: they are very dense and absorb certain radiations very well: and they can be fissioned. Therefore, there are three basic ways to detect fissile material: by observing the radiations emitted during its radioactive decay, by "X-raying" or radiographing an object to detect dense and absorptive materials, or by irradiating an object with neutrons or photons and detecting the particles emitted during fission. The first technique is often called "passive detection," because it relies on the intrinsic radioactivity of fissile material. The other techniques are called "active detection" because they use an external source of particles. This paper will focus on passive detection, since it is the preferred technique for verification purposes because of its simplicity and safety.

6.2 Weapon models

Before turning to passive detection, we must have a model of a nuclear warhead. The detailed design of nuclear weapons is secret, of course, but the general characteristics of a fission weapon are by now well known. A fission explosive can be represented by a series of concentric shells, with the fissile material on the inside. This is surrounded by a reflector, a tamper, a layer of high explosive, and some sort of case. In this paper, we explore

Table 6.1. The composition of highly-enriched uranium and weapons-grade plutonium used in this study, in percent by weight

Highly-enriched Uranium		Weapons-grade Plutonium	
Uranium-234	1.0	Plutonium-238	0.01
Uranium-235	93.5	Plutonium-239	93.5
Uranium-238	5.5	Plutonium-240	6.0
Other[a]	0.2	Plutonium-241	0.4
		Plutonium-242	0.04
		Other[a]	0.2

[a] Oxygen concentration set at 0.2 percent to give the observed (α,n) production rate from WGPu. HEU may be contaminated with uranium from reprocessed reactor fuel, thus making HEU far more radioactive

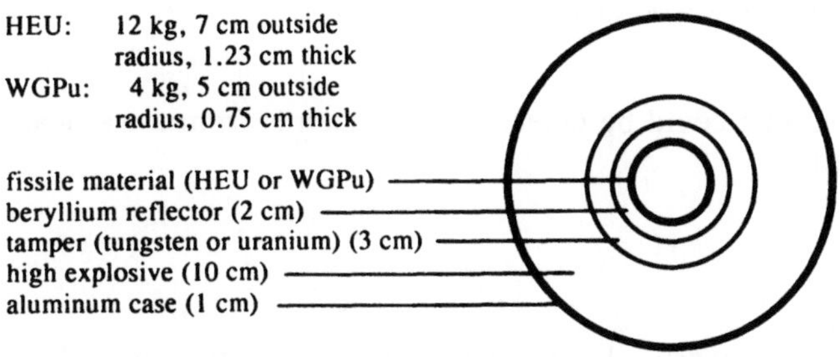

Fig. 6.1. Hypothetical weapon models used in this study

using either HEU or WGPu as the fissile material, and either tungsten or depleted uranium for the tamper, giving four hypothetical models. (Depleted uranium is 99.8 percent uranium-238.) The models are depicted in Fig. 6.1. Although we should emphasize that these models were constructed without knowledge of actual US or Soviet designs, their size and mass match those of US weapons reasonably well.

6.3 Passive Detection

All isotopes of uranium and plutonium are radioactive. The detectability of this radioactivity varies widely from isotope to isotope depending on the halflife and the types of radiations emitted during radioactive decay. The

Table 6.2. Neutron production from spontaneous fission and (α,n) reactions in HEU and WGPu, in neutrons per second per kilogramme

Isotope	Neutrons/s/kg of Isotope Spont. Fission	(α,n)[a]	Neutrons/s/kg HEU or WGPu
U-234	5.7	50.	0.56
U-235	0.30	0.012	0.29
U-238	14.	0.001	0.75
		Total HEU	1.6
Pu-238	26,000,000	220,000	260
Pu-239	22	630	610
Pu-240	910,000	2,300	55,000
Pu-241	500[b]	22	2
Pu-242	1,700,000	33	690
		Total WGPu	56,000

[a] Assuming an oxygen concentration of 0.2%
[b] Mostly from americium-241, a decay product of plutonium-241

two types of radiations that could be detected a few metres from a warhead are neutrons and photons.

6.3.1 Neutrons

Neutrons are produced primarily by spontaneous fission; that is, when isotopes of uranium or plutonium split apart on their own, without the help of an incident particle. This happens most readily in isotopes that have an even number of neutrons and protons (e.g., plutonium-238, 240, and 242). Isotopes of plutonium undergo spontaneous fission far more readily than isotopes of uranium, leading to much higher rates of neutron emission. Table 6.2 gives the rate of neutron production for each isotope, and the contribution of each isotope to the neutron production in HEU and WGPu.

Neutrons are also emitted by light elements, such as carbon and oxygen, when they absorb an alpha particle. These are called "(α,n)" reactions. Since isotopes of uranium and plutonium emit alpha particles, and since HEU and WGPu contain small amounts of light-element impurities, (α,n) reactions make a secondary contribution to neutron production in fissile material. The magnitude of this contribution is given in Table 6.2.

The neutrons produced by spontaneous fission and (α,n) reactions cause the release of additional neutrons by weapon materials. First, neutrons will multiply in the fissile material as they induce additional fissions; the

Table 6.3. The rate of neutron emission at the surface of the four hypothetical weapon designs

Weapon Model		Emission Rate at
Fissile Material	Tamper Material	Surface of Model (neutrons/second)
12 kg HEU	Tungsten	30
12 kg HEU	79 kg Depleted Uranium	1,400
4 kg WGPu	Tungsten	400,000
4 kg WGPu	52 kg Depleted Uranium	400,000

neutrons released in these fissions may also cause fissions, and so on. The shape of the fissile material determines the degree of multiplication. (In our weapon models, each spontaneous fission causes about one additional fission.) Second, some materials that may be present in weapons, such as beryllium, emit two or more neutrons when they absorb a high-energy neutron.

Neutrons are also slowed down and absorbed by surrounding materials. Neutrons lose energy by bouncing off other atoms like colliding billiard balls. This process, which is called "elastic scattering," is best accomplished with light atoms such as hydrogen or beryllium. Heavier atoms can absorb a fast neutron and emit a slower neutron and a photon; this is called "inelastic scattering." Finally, slow neutrons are absorbed by most elements, accompanied by the release of a photon or sometimes an alpha particle. These are called "(n,γ)" and "(n,α)" reactions.

Predicting neutron emissions from spontaneous fission and (α,n) reactions is easy because they depend only on the composition of the fissile material. Neutron multiplication, scattering, and absorption, on the other hand, depend on the shape and size of the fissile material and other weapon materials. Accounting for these effects is very tedious, because each neutron can undergo several reactions before it leaves the weapon or is absorbed by the weapon materials. We have used the computer program TART to predict the emission of neutrons from our hypothetical weapon models.[2] The results appear in Table 6.3.

Note that the emission rate from the surface is almost twice as great as the production rate from spontaneous fission. This is due to multiplication from fission and (n,2n) reactions. In the WGPu/tungsten model, a given neutron has only about a 22 percent chance of being absorbed in the weapon materials.

Table 6.4. The rate of the strongest photon emissions at the surface of the four hypothetical weapon designs

Weapon Model: Fissile Material	Weapon Model: Tamper Material	Emission Rate at Surface of Model (photons/second)	Photon Energy (MeV)
12 kg HEU	Tungsten	30	1.001
12 kg HEU	Depleted Uranium	100,000	1.001
4 kg WGPu	Tungsten	600	0.662
		1,000[a]	≈ 1.5
4 kg WGPu	Depleted Uranium	60,000	1.001

[a] There are about 15 neutron-induced photons per keV at an energy of about 1.5 MeV. In a low-resolution (sodium-iodine) detector, the number of counts per 70-keV channel would be about 1,000

6.3.2 Photons

Photons are released from fissile material primarily during the radioactive decay of isotopes of uranium and plutonium. Photons are also produced during fission, during the inelastic scattering and absorption of neutrons, and during the decay of radioactive isotopes produced by these reactions. Only photons with energies greater than 0.5 mega-electron-volt (MeV) or so can escape in sufficient numbers to be detectable.[3]

Unlike neutron emissions, most photon emissions occur at particular energies that are determined by the nuclear structure of the parent isotope. And unlike neutron detectors, photon detectors are available with very high energy resolution; that is, they can determine the energy of a photon with great precision. For these reasons, photon detection is best accomplished by looking for emissions in a narrow energy range, where the ratio of the signal from the weapon to the background from environmental sources will be highest. This is in contrast to neutron detection, in which neutrons of all energies are used.

A precise calculation of the number of photons emitted from the surface of our weapon models at each energy is very tedious. Our calculations included photons emitted at over 1,000 energies during the radioactive decay of 59 different isotopes. We also included a few of the strongest photon emissions from the radioactive isotopes produced by fission (delayed fission photons). The photons emitted during fission (prompt fission photons) and inelastic neutron scattering are also included, but these have a more-or-less continuous distribution of energies and are therefore less useful with high-resolution detectors.

Using TART, we calculated the fraction of photons produced in the fissile material that escape from the weapon unscattered (i.e., undegraded in energy). For the two models with a depleted-uranium tamper, we also calculated the fraction of photons produced in the tamper that emerge from the weapon unscattered. We then multiplied the strength of each photon emission by the fraction of photons of that energy that escape, and selected the strongest emissions. The results appear in Table 6.4.

In all but one case (WGPu with tungsten tamper), the strongest photon emission is at 1.001 MeV. This photon is emitted by a decay product of uranium-238 in HEU and depleted uranium. Of course, this emission is much weaker in the HEU/tungsten model since there is so little uranium-238 in HEU.

In a high-resolution photon detector, the most prominent emission from the WGPu/tungsten model is probably the 0.662 MeV photon emitted by a decay product of plutonium-241. In a low-resolution detector, the neutron-induced photons at energies of about 1.5 MeV may be more detectable.

6.4 Radiation Detection

As one moves away from the weapon models, the flux of neutrons and photons (particles/second per unit area) decreases by a factor proportional to the square of the distance. For example, the particle flux at 2 metres is four times smaller than at 1 metre: at 3 metres it is nine times smaller, and so on. At some distance, the emissions from the weapon will become undetectable because the flux will be small compared to the flux of the natural background radiation.

How far away can our weapon models be detected? To get a rough idea, let us first calculate the distance at which the signal is equal to the background, which is given by the following:

$$\frac{A\eta S}{4\pi r^2} = B \qquad (6.1)$$

where A is the area of the radiation detector (m^2) and η is its efficiency, S is the source strength (particles/sec), r is the distance from the source to the detector (m), and B is the average background rate measured by the detector (particles/sec). Solving this equation for r, we have

$$r = \left[\frac{A\eta}{4\pi B}\right]^{1/2} S^{1/2} \text{ (metres)} \qquad (6.2)$$

This equation is valid for distances of 1 to 100 meters. It should be noted that the average distance that a fission neutron travels in air before being absorbed is about 250 meters. The average distance that a 1-MeV photon travels before being scattered is 120 meters (100 m for a 0.66-MeV photon).

A thorough discussion of neutron and photon detectors is beyond the scope of this paper. Here we will consider only hand-held detectors weigh-

Table 6.5. The area, efficiency and background rates of typical hand-held and portable neutron and photon detectors

	A (m^2)	η (%)	B [a] (s^{-1})	$\left[\frac{A\eta}{4\pi B}\right]^{1/2}$	$\left[\frac{A\eta}{20\pi\sqrt{B}}\right]^{1/2}$	Energy (MeV)
Neutron						
Hand-held	0.02	5	0.1	0.028	0.0071	
Portable	0.3	20	2.	0.049	0.026	
Photon						
Hand-held	0.005	23	0.5	0.014	0.0051	0.66
		15	0.1	0.024	0.0061	1.0
		9	0.02	0.042	0.0071	1.6
Portable	0.3	40	50.	0.014	0.016	0.66
		35	50.	0.013	0.015	1.0
		30	10.	0.027	0.021	1.6

[a] For neutrons, at sea level and 54° latitude; for photons, average terrestrial values

Table 6.6. The distance in metres from the centre of the four weapon models at which the neutron and photon signals equal the background for the detectors in Table 6.5

Weapon Model: Fissile Material	Tamper Material	Distance (meters): Neutrons	Photons
12 kg HEU	Tungsten	0.2-0.3	<0.1
12 kg HEU	Depleted Uranium	1- 2	4-8
4 kg WGPu	Tungsten	20-30	0.2-0.8
4 kg WGPu	Depleted Uranium	20-30	3-6

ing about 10 kg and portable detectors weighing less than 100 kg. More massive detectors would be more sensitive, of course, but their application to treaty verification would probably be limited to fixed portals through which objects to be inspected would pass. Table 6.5 summarizes the neutron and photon detector characteristics used here; Table 6.6 gives the distance from the four weapon models at which the neutron or photon signal equals the background in the detector.

As Table 6.6 shows, the neutron signal from weapons containing WGPu is greater than the background up to distances of 20 to 30 metres. The photon signal from weapon models that use a depleted uranium tamper is greater than the background up to 3 to 8 metres. For the weapon model containing only HEU, however, the neutron and photon signals are below the background even at the surface of the weapon model.

Signals smaller than the background can be detected if the mean background rate is well known and fairly constant. One must simply wait until the signal is larger than normal statistical fluctuations in the background. The size of fluctuations in the background grows with the square root of time, while the number of particles collected grows linearly with time. This criterion can be expressed by the following equation:

$$\frac{A\eta S\tau}{4\pi r^2} \geq m\,(B\tau)^{1/2} \tag{6.3}$$

where τ is the detection time (seconds) and m is the number of standard deviations in the background that the signal must exceed before we count it as a signal. With m = 5, there is less than two chances in ten million that a chance variation in the average background could be mistaken for a signal. Using this value of m and solving for the distance, we have

$$r = \left[\frac{A\eta}{20\pi\sqrt{B}}\right]^{1/2} S^{1/2}\,\tau^{1/4} \quad \text{(metres)} \tag{6.4}$$

By substituting the values for A, η and B from Table 6.5, and the values of S from Tables 6.3 and 6.4, into the above equation, we can obtain the maximum distance at which each weapon model can be detected by using a hand-held or portable detector for a given amount of time. The results are shown in Table 6.7. For reasonable detection times (about 1 minute), neutrons from the WGPu models can be detected at distances of up to 50 metres; photons from depleted-uranium-tamper models are detectable at distances of up to 10 or 20 metres; the HEU-only model is virtually undetectable.

We should stress that in many situations the background may not be well known or constant: the distances reported in Table 6.7 should, therefore, be viewed as theoretical maxima for the weapon models and detectors under consideration.

6.5 Accuracy of Detection Analysis

We believe that our estimates of the detectability of neutron emissions from weapons that use WGPu cannot be far from the mark. First, the amount of WGPu in a warhead cannot be much larger or smaller than 4 kg, and the

Table 6.7. The maximum detection range for a given detection time for neutron and photon emissions from each weapon model

Weapon model			Distance (metres)	
FM	Tamper	Detection Time, τ	Neutrons	Photons
HEU	Tungsten	1 sec	<0.1	<0.1
		1 min	0.1-0.3	0.1-0.2
		1 hour	0.3-0.9	0.3-0.7
HEU	Depleted Uranium	1 sec	0.3-1	2- 5
		1 min	0.8-3	5-15
		1 hour	2 -8	15-40
WGPu	Tungsten	1 sec	5- 20	0.1-0.7
		1 min	15- 50	0.3-2
		1 hour	40-130	1 -5
WGPu	Depleted Uranium	1 sec	5- 20	2- 4
		1 min	15- 50	4-11
		1 hour	40-130	11-30

concentration of plutonium-240 in current warheads cannot be much larger or smaller than 6 percent. Some US warheads use "super-grade" plutonium, which is only 3 percent plutonium-240.[4] In this case, the distances in Tables 6.6 and 6.7 would be reduced by 30 percent. The US successfully tested a warhead using fuel-grade plutonium (which has about three times more plutonium-240 than WGPu), but the superpowers would have no reason to stock such weapons as long as adequate supplies of WGPu are available.[5] Making the fissile material more compact (e.g., spherical) might greatly increase neutron emissions, but the material would then be much more vulnerable to accidental detonation.

Neutrons are not readily absorbed by the materials in our weapon models, but we cannot rule out the use of neutron absorbers designed to protect against fratricide effects or antiballistic missile attack. Such materials would at most be only a few centimetres thick, and could not decrease the neutron flux by more than a factor of ten at most. A large amount of lithium deuteride, which is an effective neutron absorber, is used in thermonuclear weapons, but presumably it is not distributed around the fission trigger. (The fusion fuel in a thermonuclear weapon is contained in a physically separate component.) Absorption of fission neutrons by air is unimportant at distances of less than 100 metres.

The neutron background, which is due to cosmic rays, increases with altitude and geomagnetic latitude. Table 6.5 gives typical background values at sea-level and 54° North latitude: at 700 metres altitude the neutron flux is twice as large, and at 1,600 metres it is five times greater. Near the poles the neutron flux will also be several times greater, especially during the maximum phase of the solar activity cycle. In addition, the neutron flux near a nuclear reactor may be up to 100 times greater than the average background flux, leading to detection distances three to ten times less. (The maximum permissible dose for members of the public is 0.5 rem/year in many states. If the reactor is shielded such that unrestricted access is permitted outside the shield, then the neutron flux would be less than about 0.5 neutrons/(cm^2s). - The neutron background from reactors can be eliminated by shutting the reactor down.) These considerations are unlikely to be important except for very special detection problems.

One might suppose that detection could be improved by making a detector capable of determining the energy of the neutrons collected. Such detectors cannot readily be made efficient, however, and since about half the background flux is in the energy range of the fission neutrons, using energy-sensitive neutron detectors would not help much.

Thus, for current warheads using WGPu as the fissile material, the distances required for the detection of neutron emissions using portable detectors should not be more than a factor of two less than the values given in Tables 6.6 and 6.7. If the WGPu is in a more compact and critical configuration, these distances could be much greater.

The detectability of photon emissions, on the other hand, is subject to much greater uncertainties for two reasons. First, the rate of photon emission from a weapon is more sensitive to details of its design. To demonstrate this, let us assume, as some analysts do, that thermonuclear weapons are surrounded by a case made of depleted uranium having a surface area of about 0.5 m^2. If the case weighs 20 kg (roughly one-tenth the mass of the warhead), it would be about 0.2 centimetres thick. About one-half of the photons produced in the case would escape from the weapon, resulting in an increase of roughly 700,000 1-MeV photons per second. This is about ten times the rate calculated in Table 6.4 for a depleted-uranium tamper, and would lead to detection distances three times greater. An even greater increase in photon emissions would be caused if HEU is contaminated with uranium-232 from reprocessed reactor fuel.

Second, the photon background is less predictable than the neutron background. In three of the four weapon models, the strongest photon emission was the 1.001-MeV emission from the decay of uranium-238. The concentration of natural uranium in the soil - and therefore the background at this photon energy - can vary by more than an order of magnitude. The variation inside buildings may be even greater, depending on the concentration of uranium in building materials such as concrete. Even on ships on the open sea, the background will depend on the concentration of uranium in steel and radon in the air above the ship. Man-made sources of radiation may also create significant photon backgrounds. On ships, for example, nuclear reactors or air defence guns that use depleted-uranium bullets would increase the background at 1 MeV. Thus, the accuracy of results in Tables

6.6 and 6.7 for photon detection should be considered to be no better than a factor of three or four.

We based our analysis on detecting the strongest photon emissions. Photon detection could probably be improved somewhat by looking at the ten or twenty strongest emissions simultaneously. In low-resolution detectors, it may even be profitable to compare the total flux (including photons that have lost energy as a result of scattering) with the background. We have not done a detailed analysis of these possibilities, but it is hard to see how they could increase the distances given in Table 6.7 by more than a factor of two.

6.6 Evading Passive Detection

Passive detection is not foolproof. Some possible weapon types, such as those that contain no plutonium or depleted uranium, could not be detected by portable devices. We do not know how common such weapons are in the arsenals of the superpowers, but we presume that plutonium is the preferred material for situations in which the mass or the size of a weapon is constrained (e.g., missile warheads or artillery shells). For a given treaty regime, persons knowledgeable about weapon design would have to judge whether the covert deployment of pure-HEU warheads on certain delivery vehicles could pose a significant military threat. A ban on nuclear testing would greatly inhibit the development of new warheads designed to evade detection.

Even plutonium weapons could escape detection by one of two methods: shielding or purification. Assume, for example, that a cheater would want to be sure that a nuclear weapon could not be detected by a hand-held detector aimed at the weapon system for 1 minute from a distance of 1 metre. In our WGPu models, a neutron shielding factor of about 600 would be required to meet this criterion. A blanket of lithium-hydride 20 cm thick and weighing 300 kg would provide this much shielding. To hide the weapon model from the portable detector with a detection time of 10 minutes, 1 tonne of lithium-hydride would be required. Reducing the photon signal from a depleted uranium tamper would require comparable amounts of shielding: 600 to 1,200 kg of tungsten. Such large amounts of shielding are impractical for the normal deployment of most weapons, although covert warheads could be stored in shielded boxes or rooms during an inspection.

Neutron emissions could also be diminished by reducing the concentration of plutonium-240 in WGPu. Reducing the rate of neutron emission to an undetectable level of 100 to 1,000 per second would require plutonium-240 concentrations 500 to 5,000 times less than those in WGPu. It appears that this degree of purity is achievable using atomic vapour laser isotope separation, but only at a cost of one to several million dollars per warhead.

6.7 Notes and References

1 For a detailed discussion of the assumptions underlying these and other values in this paper, please see our forthcoming article in Science and Global Security

2 Ernest F. Plechaty and John R. Kimlinger, TARTNP: A Coupled Neutron-Photon Monte Carlo Transport Code, UCRL-50400, vol. 14, Livermore, CA: Lawrence Livermore National Laboratory, 4 July 1976

3 MeV is a unit of energy equal to $1.6 \cdot 10^{-13}$ joules. The photons released during radioactive decay have energies ranging from a few thousandths of a MeV to several MeV

4 T. B. Cochran, W. M. Arkin, R. S. Norris and M. M. Hoenig, Nuclear Weapons Databook, vol. II, U.S. Nuclear Warhead Production, Cambridge MA: Ballinger, 1987, p. 79

5 Cochran et al. (note 4), p. 24

7. Disposal of Fissile Material from Nuclear Weapons

Alfred Lecocq

7.1 Introduction

Nuclear disarmament will become irreversible when all the 50,000 warheads will have been destroyed, and when new technical barriers will be used to favour civilian development of nuclear energy, without the production of new weapon-usable materials.

It is a well-known fact that: the thorium fuel cycle has a higher proliferation-resistance than the uranium-plutonium fuel cycle; the reprocessing of the spent fuel which may lead to proliferation can be reduced by the use of the liquid fuel concept.

In this paper the proposal is offered to shift from the solid-fuel concept using the uranium-plutonium cycle, to Molten Salt Reactors with liquid fuel and the thorium-uranium fuel cycle, due to the efficiency of the latter to destroy nuclear warheads and prevent nuclear proliferation.

7.2 How to Achieve Nuclear Disarmament?

As has been shown in previous chapters the verification of arms reduction is not a simple task. The technical ways used to destroy nuclear weapons, may have a great effect on the ability to achieve verification and compliance.

The fissile material of nuclear warheads is made up of highly enriched uranium-235 or plutonium-239, the latter being called military plutonium. The former material is easily rendered harmless by simple dilution into natural or depleted uranium. For plutonium the problem is not so simple, because it is a chemical element which cannot be diluted irreversibly into another material; mechanical or physical dispersion into the oceans or by throwing into space would be expensive or insufficiently safe. The destruction by blasting under ground would also be costly and consume much energy.

To save the energy contents of the warheads, a safer and more economical way is to split the plutonium nuclei in fission reactors. It is also the best solution to make disarmament irreversible, provided that no new plutonium is produced during the splitting processes.

7.3 Recycling Plutonium in Nuclear Reactors

A halt to nuclear energy production would induce, in the next century, a large growth of fossile consumption with production of CO_2 and the well-known greenhouse effect. The disastrous effects of this phenomenon can be avoided by an intensive development of nuclear energy.[1]

The extension of nuclear industry with today's technology (LWR, HWR, HTGR or FBR) would result in a big increase of the reprocessing step; the enormous amount of "civilian" plutonium recovered - and recycled -, should be surveyed along the whole cycle from the fabrication plants to the users in the power stations. The verification needed to avoid a terrorist diversion, would also involve a big effort which cannot be reduced by denaturation, as for uranium, or by spiking with radioactive materials.[2]

The present proposal is to replace, as much as possible, the uranium-plutonium cycle by the thorium-uranium cycle, which does not produce weapon-grade materials, and to develop a new concept which opens the way to fission energy without proliferating reprocessing; this new philosophy is based on the development of *Molten Salt Reactors*.

7.4 Molten Salt Reactors (MSR)

MSRs are fission reactors using fluorides of uranium-235, plutonium or uranium-233 as fission material. These compounds have a rather high melting point (over 1000 °C); it is also easy to dissolve them in a solvent of low melting point, a mixture of lithium and beryllium fluorides. Fertile materials such as uranium fluoride - or better thorium fluoride - can be added to produce new fissile material inside the operating system (Fig.7.1).

The MSRs, in their optimized concepts, are *thermal* reactors; this means that graphite is put inside the vessel, so as to slow down the neutrons, and to have a good conversion factor of thorium into uranium-233.[3]

Two prototypes - ARE and MSRE - were operating in Oak Ridge National Laboratory; many other projects were produced there [4] or elsewhere,[5] and many nations have experts in that field: USA, USSR, UK, GDR, India, Poland, Switzerland, Japan (Fig. 7.2).[6]

Thorium resources are not localized like uranium, and it is possible to extract thorium everywhere on the earth, except in Europe: Brazil, Africa, India, Australia, Egypt, Turkey, Greenland ...

7.5 How to Burn Nuclear Weapons?

Concerning the disposal of weapon material many fissile-fertile mixtures can be considered, according to the prevailing economical or political conditions:

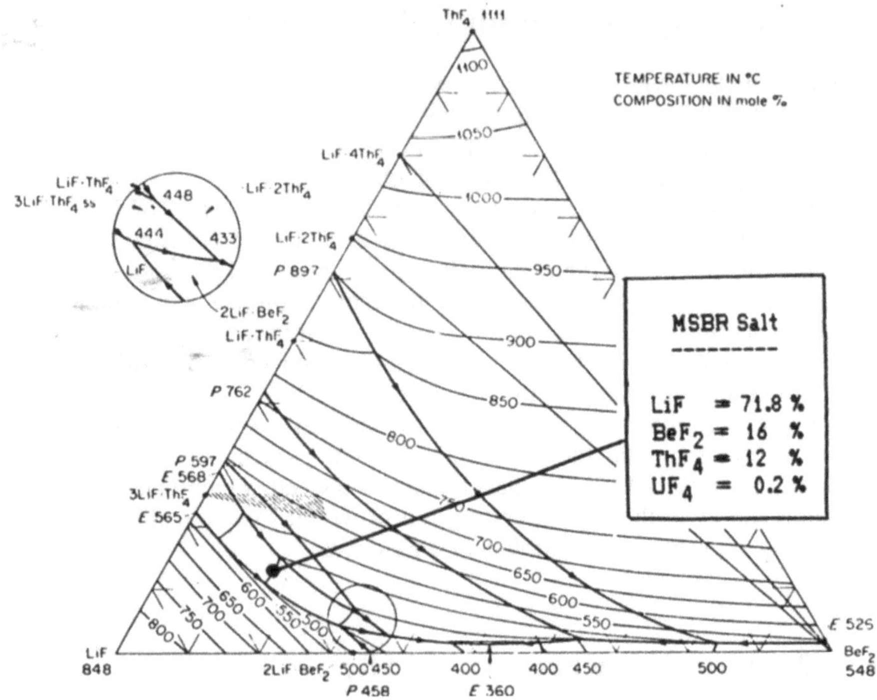

Fig. 7.1. Ternary diagram for $LiF-BeF_2-ThF_4$ [7]

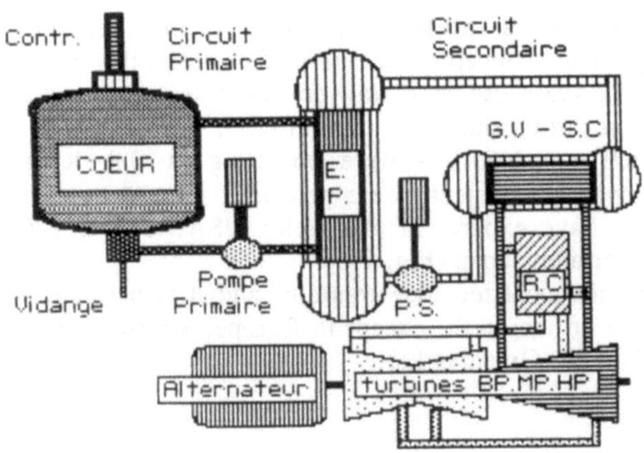

Fig. 7.2. General scheme of a 50 MW power station. E.P.: primary exchanger; P.S: secondary pump; G.V: vapour generator; S.C: preheater; R.C: reheater; turbines: BP: low pressure; MP: medium pressure; HP: high pressure

- mixture without fertile material for a swift fissioning of warheads in reactors producing solely energy or energy and tritium for fusion purposes;[8]
- mixtures with fertile materials - uranium and/or thorium - for an economical consumption of weapons in a more or less denaturated salt;[9]
- mixture with LWR plutonium prepared from spent fuel by a light dry reprocessing.[10]

All the reactors considered for this purpose, should be simple near breeder reactors, with a low power density and a very small reprocessing limited to the extraction of gaseous fission products by bubbling of helium in the primary salt.

7.6 Proliferation Advantages of MSR

The main advantages of these reactors result from the prefered use of thorium, together with the liquid state of the fuel. They concern nuclear material resources, the management of the fuel cycle, safety, radioactive wastes and their incineration, public acceptance and proliferation.

"With the thorium-uranium fuel cycle it is not possible (Fig. 7.3) to produce new weapon-useable plutonium, that is to say a material with a weak contents in isotope 238; and so, under the neutron flux, the military plutonium is polluted by the other isotopes, fissioned and finally eliminated as a weapon material." (Fig. 7.4)

"The ^{233}U produced from ^{232}Th by neutron capture, is always spiked by its 232 isotope, a high energy gamma emitter, which acts as a deterrent to any proliferation use." (Fig. 7.5)[12]

"The most important new point resulting from the liquid state of the fuel, is that the classical reprocessing method, using basically the Purex process, might disappear. Indeed, with the solid-fuel concepts, the reprocessing of spent fuel is necessary in order to recover the whole energy contents of uranium or thorium. By contrast, with the liquid fuel concepts it is possible to extract only the fission products with a big neutron cross section. This is easily done for the inert gases such as xenon-135, and it is enough to have highly efficient reactors which are near-breeders."

7.7 Cost of Destroying Nuclear Weapons

Many assessments of the price of a big size MSR, compared with the present industrial power stations (LWR), have shown that the capital cost is slightly higher for MSR, and the operating cost, rather lower. So the energy price behaves in the same order of magnitude.[13]

If we consider the small power stations, which are needed for the developing countries of the Third World, the comparison favours the MSR, because of the absence of fabrication and reprocessing cost, with their very expensive processes.

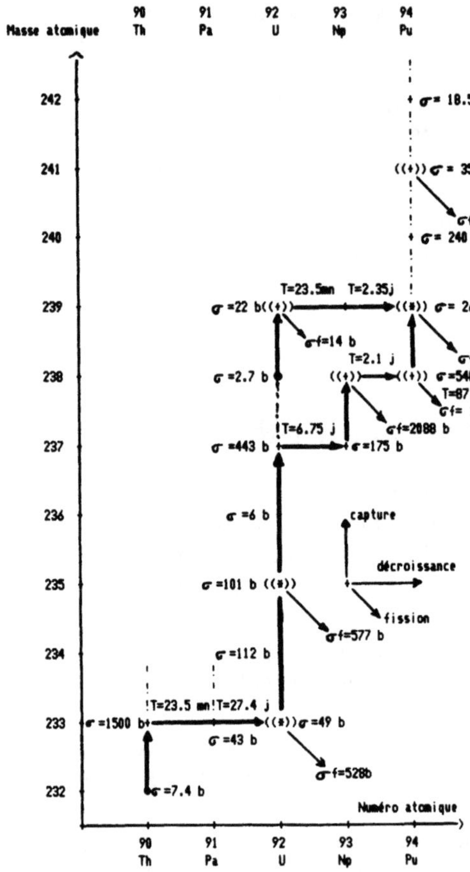

Fig. 7.3. Captures and decays from thorium-232

A rough evaluation of the price to pay for fissioning all the nuclear weapons in MSRs of various sizes, gives a value of about $200 billion, that is to say one fifth of the annual world expenses for armament.[14]

7.8 Conclusion

MSRs bring the only solution for destroying plutonium warheads, and increasing the peaceful uses of nuclear fission without producing any new nuclear weapon-usable material.[15]

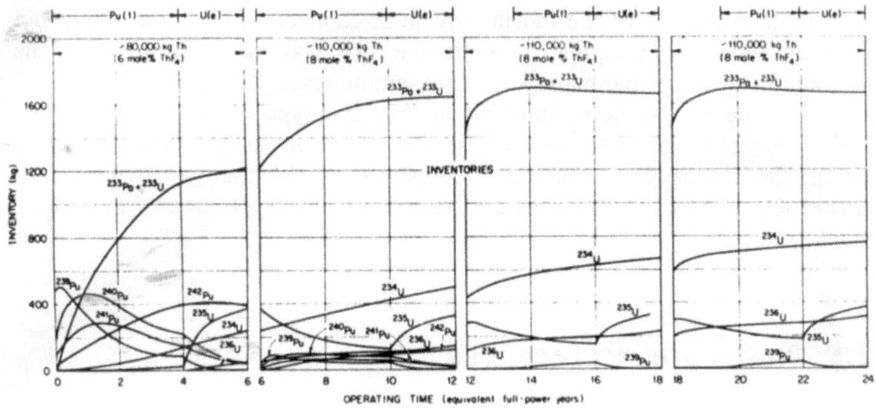

Fig. 7.4. Principle nuclide inventories for MSR with 8 mole% ThF_4 (LWR plutonium and enriched uranium feed) [11]

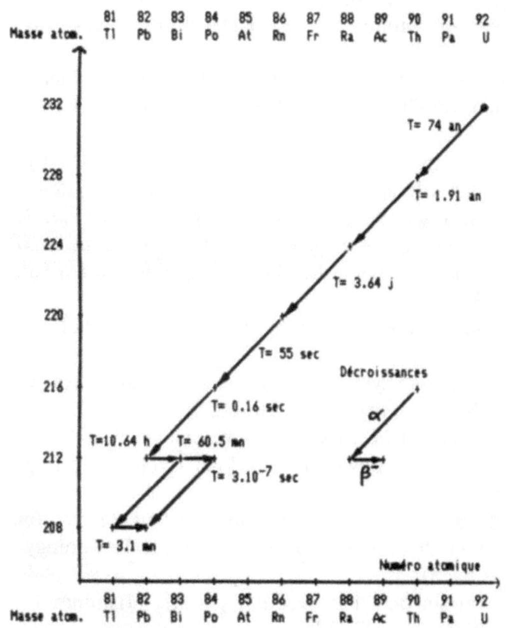

Fig. 7.5. Decay of uranium-232

The development of the thorium cycle in connection with the liquid fuel concept makes it possible to limit the classical reprocessing and the plutonium handling, the main ways to diversion and terrorist threat.[16]

The technical problems inherent in this new type of reactor have been solved 20 years ago. The price to pay for achieving nuclear disarmament is within the means of international cooperation. The main difficulty is to ensure the will of the superpowers on a common target for the welfare of mankind.

7.9 Notes and References

1 A. Lecocq, Effet de serre et Réacteurs à Sels Fondus, Coll. Avenir de l'Energie et Perspectives de la Fission, ISIB, Bruxelles, 16-17 Nov. 1988

2 D. Albright and H. Feiveson, Plutonium recycle: An Unnecessary Threat to International Security, Princeton, July 1986

3 M. W. Rosenthal, R. C. Robertson, R. B. Bridges, Molten Salt Reactor - History, Status and Potential, Fuel Appl. and Techn., vol. 8, pp. 107-117, Febr. 1970

4 H. G. MacPherson, The Molten Salt Adventure, Nuclear Science and Engineering, vol. 90, pp. 374-380, 1985

5 M. Hery and A. Lecocq, Results of and Prospects for Studies on Molten Salt Nuclear Reactors, Proc. 1st Int. Symp. Molten Salt Chemistry and Technology, p. 423, Kyoto, 1983

6 D. D. Sood, Molten Salt Reactor Concept, Board of Research in Nuclear Science, Bombay, 1980; V. M. Novikov, Molten salt advanced and emerging nuclear systems: flexible energy option, ICENES 3 - Helsinki 1983, Atomkernenergie, vol. 45, no. 1, pp. 45-49, 1984; K. Furukawa and A. Lecocq, Preliminary examination on the Next Generation Nuclear Reactors in comparison with the Small Thorium Molten-Salt Reactor, Tokai Univ., December 1988

7 R. E. Thoma, H. Insley, H. A. Friedman and D. F. Weaver, J. Phys. Chem., vol. 64, pp. 865-870, 1960

8 V. L. Blinkin and V. M. Novikov, Symbiotic system of a fusion and a fission reactor with a very simple fuel reprocessing, Nuclear Fusion, vol. 18, no. 7, 1978

9 L. F. C. Reichle, MSR (Pu Converters) and MSBRs in Commercial Nuclear Power Systems, Proc. 1977 Salzburg Conf. Nuclear Power and Its Fuel Cycle, vol. 1, p. 531, IAEA, Vienna, 1977

10 J. R. Engel, W. A. Rhoades, W. R. Grimes, J.F. Dearing, Molten Salt Reactor for efficient nuclear fuel utilisation without Plutonium separation, Nuclear Technology, vol. 46, pp. 30-43, 1979 (ORNL-TM 6413)

11 A. M. Perry and H. F. Bauman, Démarrage de filière avec PuF_3. Caractéristiques des RSF de première génération, ORNL 4676, 1971

12 J. R. Engel et al., Conceptual Design Characteristics of a Denatured Molten Salt Reactor with Once-Through Fueling, ORNL-TM 7207, July 1980

13 H. G. MacPherson, Memorandum on Molten Salt Breeder Reactor, ORAU, Nov. 1976

14 J. H. Brunn and A. Lecocq, Propositions de techniques de réduction de la prolifération nucléaire, Workshop 12, Ways Out of the Arms Race, Hamburg, 14-16 Nov. 1986

15 A. M. Weinberg, On the Development of Diversion-proof Nuclear Breeder Reactors, Pugwash, Preventing the Spread of Nuclear Weapon, 1969

16 P. Leventhal and Y. Alexander, Preventing Nuclear Terrorism, Lexington: Lexington, 1987

8. The IAEA's Controls on Fissile Material: Limits and Successes

David A.V. Fischer

8.1 The Aim of Safeguards: Verification, not Control

Let us start by dispelling some misconceptions about the purpose of International Atomic Energy Agency (IAEA) safeguards (such as that reflected in the title suggested for this paper). The IAEA's safeguards are far from being "controls" in the English sense of the word. Their aim is to verify and certify that safeguarded plant and material is being used solely for "peaceful" i.e. for civilian purposes. Safeguards should be seen as a confidence-building measure, not as a police operation.

As a rule, a government gives its "peaceful use"[1] undertaking freely and voluntarily, often to assure other states about its nuclear activities. The normal presumption is that a state will abide by its undertaking to ensure that its programme is peaceful. However, to be credible, safeguards must be able to detect promptly any non-compliance. An inadequate system might be worse than no system at all: it might lull nations into a false sense of confidence. Hence the technical role of IAEA safeguards is defined as "...the *timely* detection of diversion of *significant* quantities of nuclear material from peaceful nuclear activities to the manufacture of nuclear weapons or of other nuclear explosive devices or *for purposes unknown*, and *deterrence* of such diversion by the *risk* of early detection."[2]

Even the assumption that safeguards should have a deterrent purpose has given way in recent years to emphasis on the positive assurance they should provide. This is no more than realistic. If ever a powerful nation like the FRG or Japan or, for that matter, Canada or Sweden, were to take the fateful decision to "go nuclear" - which is highly improbable - it would not be deterred from doing so by IAEA safeguards. It would surely not try to "go nuclear" by clandestinely diverting relatively inconsequential amounts of fissile material from its civilian nuclear fuel cycle. It would withdraw from the nuclear non-proliferation treaty (NPT) thus terminating IAEA safeguards, and it would plunge ahead into a crash nuclear weapon and missile programme.

8.2 Inherent Constraints in a World of Nation States

A second point to be stressed is that until there is world government, which is not imminent, any organization which monitors an arms control or disarmament agreement is likely to be subject to constraints and to encounter difficulties similar to those that affect the IAEA.[3] Hence the lessons one can draw from the IAEA's experience are valuable, indeed unique.

8.3 Incomplete Coverage

If limited authority constrains the IAEA, a second, perhaps even more important, constraint is that the coverage of IAEA safeguards is incomplete. A glance at the map will show the vast areas of the world where nuclear weapons are being or could soon be made. For instance, in the most powerful and in the most populous countries of the world.

In the five states that have declared nuclear arsenals, the USA, the USSR, China, UK and France, the IAEA applies safeguards only to a handful of selected plants and clearly not to verify that the nuclear programmes of those states are wholly peaceful! In the six so-called "threshold states" - not parties to the NPT - namely, India, Pakistan, Israel, South Africa, Argentina and Brazil, there are unsafeguarded plants that can make nuclear weapon material.

India has tested a nuclear explosive; by all accounts Israel has a formidable stock of nuclear weapons; Pakistan and South Africa have said, or implied that they could make them, and the same may soon be true of Argentina and Brazil though they are much less likely to do so.

Why should incomplete coverage be regarded as a constraint? In the threshold states it makes the application of safeguards technically more difficult and politically less meaningful. In the five weapon states the absence of safeguards weakens the credibility and acceptability of the non-proliferation regime.

The nuclear weapon states (NWS) constantly proclaim that nuclear weapons are essential for their security. One can hardly blame India, Pakistan or Israel if they make the same argument. But keeping in check the number of nuclear weapons states is an essential condition for reversing the nuclear arms race between those states that already have them, and for moving towards a world in which the threshold states as well as every other nation would be more secure.

8.4 Other Limitations

A third constraint is that the IAEA's inspectors may not roam around freely in search of unreported plant and material, they may usually visit only those plants or stores where the government concerned has declared nuclear material to be located.

The IAEA must also secure the agreement of the state concerned to the inspectors it wishes to assign to that state. This procedure reflects the sovereign right of states to refuse admission to a person they do not like. States are unlikely to surrender this right. A similar procedure is prescribed in the INF Treaty (and the USA has already rejected certain Soviet inspectors) and is also foreseen in the draft convention banning the manufacture of chemical weapons (the CWC).

But in the IAEA's experience states do not simply use this right to reject Mr. Smith or Mr. Semenov or Mr. Subramanian. They use it in a way that lowers the efficiency of safeguards operation and introduces unwanted political factors. Many states simply refuse to accept in advance any inspectors of particular nationalities or from the other "bloc".

Finally, the IAEA cannot divine the intentions of any state.[4] It works in the here and now. It can verify that the state has not, or is not diverting nuclear material but it cannot presume to guess what a particular government has in mind for the future. Sometimes the overall nuclear programme of a state may give other nations a clue to its intentions but this is not a judgement that the IAEA can make.[5]

Time may alleviate several of these problems. The coverage of safeguards is continuing to expand. Spain's recent adhesion to the NPT closed an important gap; Saudi Arabia and Bahrein have recently acceded, this has a symbolic as well as a substantial importance in the tense Middle East. And though the IAEA's inspectors may not be allowed to roam around freely in any country, the IAEA does have the right to call for special inspections at locations not previously defined if it is dissatisfied with the information it is getting from a particular government.[6] It has never signalled the need to do so. Recently, too, the Director General of the IAEA has proposed a radical reform of the process of designating inspectors. The USA and the USSR and some other countries have accepted the reform, in whole or in part.

8.5 How Safeguards Work

Briefly, safeguards seek to account, as meticulously as possible, for all nuclear material as it moves through a national fuel cycle or across national frontiers.

At the start of an accounting period the manager of the plant takes stock (makes a "physical inventory") of the amount and nature (e.g. the isotopic composition) of all nuclear material in each previously designated section of his plant. The manager of the plant must keep meticulous records (for the sake of safe and economic operation as well as safeguards) of every movement of and change in material. Each month the manager reports to the IAEA (via the national or regional authority) every movement/change in each section of his plant. IAEA inspectors examine the plant's records and verify that they tally with the reports sent to IAEA.

By adding to the original physical inventory all reported inputs and production of nuclear material, and by subtracting all reported outputs, consumption and losses, the manager and the IAEA arrive at a "book inventory".

At the end of the accounting period the plant manager takes another physical inventory of the material actually present in each section of the plant. The IAEA's inspector checks this physical inventory and compares it with the calculated book inventory. Any difference that he finds is recorded as material unaccounted for ("MUF"). If the difference, i.e. the MUF, is significant a satisfactory explanation for the MUF must be found, if necessary by special additional inspections, not necessarily limited to the plant.

This system of "material accountancy" is made more efficient and more cost-effective by the use of instruments (e.g. video recorders which record all movements of material in sensitive areas of the plant) and by "containment" (e.g. the use of tamper-resistant seals).

It should be emphasized that the IAEA does not verify a particular batch of nuclear material only once or twice; each batch is verified virtually every time it changes location or form; as many as thirty to forty times during its "life" in the fuel cycle. Moreover, the IAEA's assessment of what is happening at a particular plant is not made in isolation on the basis of a single verified report. It takes into account a wealth of additional information - verified reports from other plants sending or receiving material from the plant in question, the reports of previous accounting periods, information coming from surveillance equipment and containment measures, information about the operation of the plant, and, of course, the direct observations of the inspector.

The IAEA aims to be able by *material accountancy* alone to detect the diversion of a significant quantity of nuclear material (roughly the amount of material needed by a country to make its first nuclear weapon) with a level of confidence of 95%.[7]

8.6 The Size of the Operation

The IAEA has earmarked about $52.5 million for safeguards in 1989 (cf. about $44.5 million for all its other programmes). The IAEA has about 200 inspectors and a back-up safeguards staff of about 90 professionals. This compares with 800 inspectors that, it is reported, will be needed to verify the INF Treaty and a US outlay of $200 million in the first year and $100 million a year afterwards.

To quote a few more figures, in 1987 the IAEA carried out 2,133 inspections in 52 states (the frequency of inspection ranges from once a year at a small research reactor to full-time around the clock inspection at a large plant handling large quantities of e.g. plutonium).

The rate of growth of the IAEA's responsibilities can be gauged from the amount of plutonium under safeguards. In 1977 the total was 34 tonnes, by 1987 it had risen to 181 tonnes (including 8.8 tonnes of separated plutonium).[8] Within the next five to seven years it is expected to reach 430 tonnes.

8.7 How Effective are IAEA Safeguards?

No international verification system can hope to be 100% technically effective. The best that one can expect is that the system will provide sufficient assurance that states are complying with their agreements. And what is "sufficient assurance" will differ according to who is making the judgement.

The 35 states that sit on the IAEA's Board of Governors make an analysis each year of the results of the previous year's operation. They base this analysis on a very detailed technical report, prepared largely by a separate evaluation unit in the Safeguards Department.[9] The report highlights all the difficulties that the safeguards operation has encountered, the cases when its technical goals could not be met and the reasons for failing to meet them. The report also suggests the remedial action that should be taken by states, by plant managers and by the IAEA.

The first report came out in the mid 1970s. Each year since then, and on the basis of the report, the IAEA has found it "reasonable to conclude that nuclear material under [IAEA] safeguards in the previous year remained in peaceful nuclear activities or was otherwise adequately accounted for".[10]

The last NPT review conference in 1985 came to a similar but even more positive conclusion about the effectiveness and value of IAEA safeguards.[11]

The governments of all the leading powers and many others have thus repeatedly given safeguards a clean bill of health.[12] The same cannot be said of the media. The annual safeguards report almost always leaks to the press, is usually misunderstood and often misinterpreted.[13]

Another gauge of safeguards effectiveness is the ability of the operation to detect anomalies - the wrong figure in an accounting report, a failed camera, a broken seal, and to remedy or discover the explanation for such anomalies. The IAEA detects about 250-300 anomalies a year - 290 in 1987 -, all of which have since been satisfactorily clarified.

The IAEA has been applying safeguards since the early 1960s. In the 26 years since its first inspection it has discovered no case where the government has diverted or attempted to divert safeguarded material or misused safeguarded plant. What is more surprising, in a world where little love is lost between nations, no country has ever alleged that another, even its inveterate adversary, had misused a safeguarded item.[14] There have, of course, been many allegations that country X was clandestinely building an *unsafeguarded* plant or producing *unsafeguarded* nuclear weapon material but there have been no allegations of breach of safeguards agreements. The danger of proliferation, vertical or horizontal, begins where safeguards end.

Finally, it is my opinion that if ever a state did decide to divert it would see to it that it was not caught red-handed, for instance, caught cooking the books of material accountancy. It would simply make it impossible for the IAEA to apply effective safeguards.

This seems to be precisely the type of situation foreseen in the standard NPT safeguards agreement. If the IAEA should suspect that there has been a diversion it is not called upon to prove it; this might be a very difficult task, for instance if a state refused to admit IAEA inspectors. If the IAEA "...is *not able to verify* that there has been no diversion..." the IAEA's Board of Governors may sound the international alarm and invoke the other sanctions foreseen in the IAEA's Statute.[15]

8.8 What Have IAEA Safeguards Achieved?

This question cannot be answered in isolation; it must be examined in the broader context of the NPT.

Dire forecasts were being made in the 50s and 60s about the inevitable spread of nuclear weapons. A man as percipient as President Kennedy foresaw a world of as many as 25 NWS by the mid 1970s - and he dreaded the prospect. Until then the barrier which prevented the spread of the bomb was technological, most states simply did not have and could not get the hardware, know-how and materials needed to "go nuclear". But as soon as they could make the bomb, they did (with one notable exception, Canada). It was generally assumed that this pattern would continue as more and more states mastered nuclear technology.

It is now estimated that as many as 35 to 40 states could make nuclear weapons if they set their minds to it and disregarded their treaty obligations. Why, 26 years after Kennedy's nightmare, are there only five official NWS and perhaps another three or four waiting in the wings?

The answer is, of course, that most nations have decided that they are better off without the bomb. When a state has so decided, the NPT and regional non-nuclear weapon treaties offers it the way to ratify the decision, to make it public and internationally binding and thus foster international confidence. IAEA safeguards provide it with the means of assuring other states that it is abiding by its decision. The assurances so provided has a feed-back effect, reducing the incentive of other states to acquire nuclear weapons or at least to keep the option open. In this way IAEA safeguards have made a contribution to international security that is now widely recognized.

This contribution can be measured in another way. 135 states have now joined the NPT, making it the most widely accepted of all the arms control and disarmament treaties. The entire industrial "North" has thus agreed to abide by it. In fact all non-weapon states in Eastern and Western Europe, all in North America, in the Far East, South-East Asia and Australasia and most states in other regions have agreed to accept IAEA safeguards on all their peaceful nuclear activities. As a result 95% of the nuclear material outside the 5 weapon states is now under IAEA safeguards. The remaining 5% in the six threshold states is of course, a major concern, as is the far greater amount in the NWS.

IAEA safeguards also provide the essential framework of international nuclear commerce. In the 50s and 60s some Western states provided nuclear training and made several critical nuclear exports without safeguards. We are now paying the price - these actions did much to help the six threshold countries get where they were in 1988.

Today there is virtually no country that would export nuclear hardware or material without requiring that it be placed under IAEA safeguards. In other words without those safeguards, there would be little or no nuclear commerce.[16] Perhaps those who consider anything nuclear as tainted by original sin would prefer it that way. But for Western Europe that already gets one third of its electricity from nuclear power, and for Japan that gets about a quarter, a collapse of nuclear trade would be a disaster today.

Historians may say that the greatest achievement of IAEA safeguards was not in the field in which they operate. It was in pioneering the first and until now the only international system of verification by regular on-site inspection. In this way nations in all parts of the world took part in constructing and committed themselves to accept a system which would open the doors of their nuclear plants to foreign inspectors. This process also helped to pave the way to wider measures of disarmament, for instance INF and the proposed CWC both of which depend critically on on-site inspection.[17] To take a particular example, the safeguards that the IAEA applies to selected plants in the five nuclear weapon states obviously do not prevent proliferation but they may have helped to inure some of those states to the novel experience of international inspection. In this sense, Soviet acceptance of on-site IAEA inspection in 1982 and China's recent conclusion of a safeguards agreement with the IAEA can be seen as breakthroughs.[18] They also signify that the IAEA is (or in the case of China will soon be) applying safeguards in every state that has a nuclear reactor or other significant nuclear plant - though not by any means to all the nuclear plants in the threshold and weapon states.

8.9 Risks and Opportunities for IAEA Safeguards

Today the great majority of the IAEA's safeguards agreements stem from the commitments that governments undertook in ratifying the NPT. Those agreements would lapse if the NPT were to terminate. In 1995 the parties to the NPT will decide how long it should be prolonged. Every option will be open. For instance the Treaty could be brought to an end a month later. This is highly improbable. If, however, it did happen IAEA safeguards would, in the worst case continue to apply only in those countries which had accepted them by 1970 and a handful of other developing countries that have since imported nuclear plant subject to "fall-back" safeguards. In several respects the situation might be worse than in 1970. Many more states would have unsafeguarded nuclear plants.[19]

Uncertainty will prevail about the future of the international nuclear non-proliferation regime including IAEA safeguards until we know the outcome of the extension conference.

The prospects for that conference and for the regime could also be darkened if any of the threshold states were to demonstrate by a nuclear test or by official proclamation that it had launched a nuclear weapon programme. We have seen that the Indian test in 1974 was widely, if unjustly, regarded as a severe blow to the NPT and to IAEA safeguards. Moreover a nuclear test on the Indian sub-continent today would probably precipitate an Indo-Pakistani nuclear arms race. A test by South Africa might induce other African states to withdraw from the NPT.

The Middle East poses a double uncertainty. No Arab state is likely to be able to make a nuclear weapon during this century. Israel might nevertheless be tempted to go public about its reputed nuclear arsenal in order to deter any thinking in Syria or Iraq about launching missiles with gas-filled warheads against Israeli targets.

The second risk lies in the US policy of defending Israel, no matter at what the cost, against hostile actions taken in international bodies and meetings. The USA has threatened to withdraw from the IAEA if the Arab states again succeed in depriving Israel of any of its rights as a member of the IAEA - as they did, temporarily, after the Israeli attack on the Iraqi reactor.[20]

The future may also offer new roles for the IAEA's safeguards. Sweden has proposed that they should be gradually be extended to cover the entire civilian nuclear programmes of the five nuclear weapon states (who have not betrayed much enthusiasm for the idea). Such an extension would help to palliate one of the most criticized aspects of the NPT: the privileged status it accords to those five weapon states.

The extension proposed by Sweden would also prepare the ground for the oft-suggested freeze on the production of nuclear material for nuclear weapons. If the NWS were to agree to such a freeze - not that there is much sign of their doing so today - the IAEA could easily and quickly extend its safeguards to cover the output of the few remaining purely military reprocessing and enrichment plants and tritium producing reactors. The entire fuel cycle of the NWS would then become "civilian" and operate under safeguards in the same way as the fuel cycles of NPT NNWS.

Finally, if the superpowers were to reach agreement on START, a considerable amount of plutonium and enriched uranium might become surplus to military requirements. This could be stored under IAEA safeguards, or possibly used for generating electric power in safeguarded civilian plants. This may seem far-fetched today. Yet it was precisely for the purpose of drawing down the nuclear stockpiles of the weapon states that on 8 December 1953 President Eisenhower proposed that an International Atomic Energy Agency should be created.

8.10 Notes and References

1 The Nuclear Non-Proliferation Treaty (NPT), which came into force in 1970, requires the 132 non-nuclear weapon states that have joined it to forswear nuclear weapons and other nuclear explosive devices. It permits them to make use of nuclear energy for military *non-explosive* uses, for instance as the propulsion units for submarines and other naval craft. Until recently no non-nuclear weapon state had made use of this option. It now appears that Canada may do so

2 IAEA document INFCIRC/153 paragraph 28 (emphasis added). This document sets forth the standard safeguards agreement between the IAEA and NPT NNWS. It will be noted that if one is to apply safeguards systematically and evenly, it will be necessary to quantify three terms:

What is meant by *timely* detection?

What is meant by a *significant* quantity?

What should be the *risk* of detection?

It will also be noted that it is a breach of the agreement to divert nuclear material for *purposes unknown* and not only for nuclear weapons or other nuclear explosive devices

3 Indeed an organization set up to verify the non-production of CW (chemical warfare) agents is likely to have greater difficulties in ensuring effective verification than one that verifies the non-production of nuclear weapons (see note 7 below)

4 For instance when Israel bombed the Iraqi TAMMUZ reactor in 1981 the Israeli government alleged that Iraq planned to use the reactor to make plutonium for a bomb and then to denounce the NPT. The IAEA affirmed that Iraq had complied with her safeguards obligations but the IAEA could not, of course, make any statement about the Iraqi government's intentions

5 The exceptionally large power of Iraq's research reactor (supplied by France), the fact that it was to use highly enriched uranium, Iraq's purchase of a "hot-cell" from Italy (a facility which would have enabled Iraqi scientists to experiment with the separation of plutonium) were among the factors that aroused Israeli suspicions.

In an ideal world all production of nuclear weapon material, i.e. plutonium and highly enriched uranium, might have been prohibited - and President Carter's administration would have liked to prohibit it in non-nuclear weapon states. But even the most visionary of all proposals so far presented for eliminating nuclear weapons and their production did not go so far. This was the "Baruch" plan which the Americans put forward in 1946 and dropped a few years later. The Baruch plan would have permitted the production of fissile materials but put such production in the hands of a UN Agency which would have owned and operated all sensitive nuclear plants throughout the world.

Twenty years later the Nuclear Non-Proliferation Treaty foresaw that the non-weapon states would be called on to place their entire nuclear industries under the safeguards of the IAEA while those of the weapon states would be exempt. Non-weapon states like FRG, Japan and even the Netherlands and Belgium would have flatly rejected an

additional discrimination which would have permitted the USA, the USSR, UK, France and China to produce fissile material but denied it to the non-weapon states.
So the NPT and IAEA safeguards leave states free to produce fissile material, *under safeguards*, and it is only the diversion of material to weapons or to other nuclear explosive devices or for purposes unknown that is prohibited. Similarly the draft Chemical Weapons Convention will permit states to manufacture, under safeguards, the key precursors of chemical warfare agents and to build pesticide factories that could potentially be misused. It is only the production of the weapons and their contents that will be monitored

6 The IAEA's system does not make provision for "challenge inspections" of the type foreseen in the Treaty for the Prohibition of Nuclear Weapons in Latin America (a provision that is still a dead letter) and which are now proposed for the CWC. However after consultations and "in agreement with the State" the IAEA "may obtain access to information or locations in addition to the access ... specified for ad hoc and routine inspections". No limitation is placed on the access that the IAEA might request for such special inspections. If the state concerned were to refuse the IAEA's request the Board might find that the IAEA was no longer able to verify that there had been no diversion of safeguarded nuclear material (see note 15 below) in other words sound the alarm. (The provisions in question are in IAEA Document INFCIRC/153 Corr. paras. 73 and 77). The IAEA has not made use of the right to carry out special inspections, presumably because it has not felt the need to do so

7 Clearly this meticulous material accountancy would be very difficult to apply in, for instance, verifying a Chemical Weapons Convention.
The IAEA deals in practice with two elements only, uranium and plutonium, and with two isotopes of uranium. Because of the radiation they emit, these elements lend themselves to precise measurement and are relatively easily traced. The amounts of material used in the nuclear fuel cycle are relatively modest. The nuclear industry still quite new and it has been possible to keep track of all nuclear plants and much of the nuclear material outside the NWS. The sensitive nuclear plants, i.e. those that produce nuclear weapon material, are confined to a dozen countries outside the NWS (those under safeguards operate in only six NNWS).
The chemical industry, on the other hand is centuries old, deals with very large quantities of material and a great variety of chemicals that can be misused, a variety that is constantly increasing. Plants capable of making chemical warfare agents are spread around the globe. Hence it seems that a CWC verification system will have to rely more on qualitative than on quantative methods of verification, on spot checks, unannounced inspections and "challenge inspections"

8 The figures do not include plutonium under safeguards pursuant to the agreements with NWS that have voluntarily offered to place some of their plants under safeguards. In 1987 such plutonium (all in irradiated fuel) amounted to 43.2 tonnes (Sources: IAEA Annual Reports for 1981, p. 63, and for 1987, pp. 76-77)

9 In their wisdom, the governments which constitute the Board insist that the "Safeguards Implementation Report" be treated as a classified document. A copy is sent to the government of each member of the IAEA; hence perhaps the inevitable leak and misinterpretation. Though largely written in the opaque jargon of the discipline it is an extremely useful analysis and can be commended to any other organization that monitors similar operations, with the rider that it should be written in a way that the media can understand

10 There were two cases in the late 1970s involving Pakistan and India where the IAEA could not reach this conclusion. Both were subsequently resolved. See: David Fischer and Paul Szasz, Safeguarding the Atom, A Critical Appraisal, Jozef Goldblat (ed.), London and Philadelphia: SIPRI, Taylor and Francis, 1985, pp. 16-17

11 The conference adopted the following text, by consensus:

"The Conference expresses the conviction that IAEA safeguards provide assurance that States are complying with their undertakings and assist States in demonstrating this compliance. They thereby promote further confidence among States and being a fundamental element of the Treaty, help to strengthen their collective security. IAEA safeguards play a key role in preventing the proliferation of nuclear weapons and other nuclear explosive devices. Unsafeguarded nuclear activities in non-nuclear-weapon States [not party to the Treaty] pose serious proliferation danger."

12 After the Indian nuclear test explosion at Pokharan in 1974, it was clear, however, that several governments of industrialized countries considered that while IAEA safeguards might be able to *detect* a diversion, neither they nor the NPT offered enough security to *prevent* a future diversion. These governments then imposed strict controls on exports (to NPT parties and non-parties alike) designed chiefly to limit the export of reprocessing and enrichment technologies.

This decision was triply ironical. Firstly IAEA safeguards are not designed to prevent diversions. Secondly, the Pokharan test was not the result of a diversion: for the test, the Indian government used plutonium from an unsafeguarded reactor (the reactor and its fuel had been supplied by Canada under Canadian safeguards which fell away as soon as India was able to replace the Canadian fuel with fuel made in India). Thirdly, India had not joined the NPT and had made it clear that it had no intention of doing so. In short, neither the NPT nor IAEA Safeguards were being judged and found wanting by the Indian test.

Nevertheless, the test was widely seen as a severe blow to both the Treaty and the safeguards system.

The export controls that the "North" imposed were also difficult to reconcile with Article IV of the NPT which promised all parties "...the fullest possible exchange of equipment, materials and scientific and technological information for the peaceful uses of nuclear energy."

13 One sees such misinterpretations particularly in papers or journals that tend to be against nuclear power. Perhaps they wish to demonstrate that nuclear energy cannot be effectively safeguarded and should therefore be abolished. One wonders whether they would try the same tack with the chemical industry

14 After bombing the Iraqi TAMMUZ reactor in 1981 the Israeli government claimed that Iraq had planned to misuse the reactor so as to produce plutonium for a military programme. The evidence adduced for such a plan was chiefly the strange assortment of nuclear fuel and plant that the Iraqi government had acquired and bellicose statements by Iraqi leaders (see note 5 above)

15 IAEA document INFCIRC/153 Corr. paras. 18 and 19

16 Outside the Common Market which has its own EURATOM system of safeguards

17 This point was made by Soviet observers when the USSR announced its willingness to conclude a safeguards agreement with the IAEA - see note 18 below. In July 1982 the Soviet journal of International Affairs published an article by R. Zheleznov on "Monitoring Arms Limitation Measures". Zheleznov noted that: "... The positive significance of the IAEA safeguards system lies in the fact that it can in many ways provide a prototype for organizing control in other areas of limiting arms, especially nuclear arms ...", International Affairs, Moscow, pp. 75-84, July 1982

18 The USSR announced in June 1982 that it was prepared to accept IAEA safeguards on certain of its nuclear plants. The safeguards agreement between the IAEA and the USSR came into force in June 1985

19 The picture would be complex. EURATOM safeguards would continue in EEC states but those safeguards do not necessarily foreclose the weapon option. IAEA safeguards would continue in those countries that had ratified the Latin American and South

Pacific nuclear-weapon-free-zone treaties (Argentina, Brazil, Chile and Cuba have not effectively ratified). In many other cases suspended agreements would again apply or "fall-back" safeguards would come into operation. But there would be many gaps

20 A recent decision by Congress formalizes the US position. It makes it mandatory for the USA to withdraw from any UN agency if Israel's rights of membership are curtailed and to continue non-participation until the rights in question are restored. Interestingly, exceptions are made precluding withdrawal from the Security Council of the UN *and from the IAEA's safeguards activities*

Part III

Verification of a Nuclear Test Ban

9. Can Nuclear Weapons Be Developed Without Full Testing?

Theodore B. Taylor

Confident development and stockpiling of nuclear weapons may or may not require full yield testing, depending on the nature of the weapons, the prior experience and technical capabilities of the developers, and the degree of assurance required that the weapons will perform as intended.

9.1 Pure Fission Weapons

Nuclear weapons that derive all their energy from fission can now be developed with high confidence, without any nuclear tests. The extent to which the performance of such weapons can be pushed towards basic limits of yields, weights, and efficiency of use of fissile materials (plutonium or highly enriched uranium), without nuclear testing, depends on the extent to which the developers make use of relevant development tools that have become widely accessible since the early days of this technology. These include published technical data (such as critical masses, equations of state, neutron cross sections); accessible procedures and equipment for calculation of weapon performance; and techniques for laboratory and field testing of weapon components and assembled weapons in which ordinary uranium has been used as a substitute for fissile materials.

Confidence in the performance of pure fission weapons can be increased by "zero yield" testing. This approach involves testing implosion systems with smaller quantities of fissile core materials than needed for the full design yield. This can be done by producing a fission yield high enough to be observable with radiation detectors close to the device, but corresponding to much less energy than released by the implosion system's chemical explosive. Such extremely low energy releases require only about 25 generations of a fission chain reaction, compared with something like 45 generations to build-up to a yield equivalent to a few tons or more of high explosive.

Such tests require much sophistication, especially if it is important that errors in design do not cause the yield to be many orders of magnitude greater than intended. This risk can be minimized by planning several tests, starting with one that involves so little fissile material that there is no possibility of a nuclear explosion, but there is likely to be observable multiplication of neutrons emitted by a source within or close to the fissile materials.

Extremely low yield nuclear tests are typically more stringent tests of the overall behaviour of a weapon assembly system than a full yield test, since imperfections in the weapon have more effect on its performance at very low yield than at high yield. Such tests can therefore be useful to countries with considerable experience with nuclear weapons and their testing. But they can also be useful to countries starting nuclear weapon development, if they want to test without detection.

9.2 Boosted Fission Weapons

The yields of fission weapons with low efficiencies, in terms of the fraction of the fissile material that undergoes fission in the explosion, can be greatly increased by incorporating small quantities of tritium and deuterium that react and release high energy neutrons into the fissioning material while the chain reaction is still rapidly multiplying the numbers of neutrons present. Increases in yield by this process, called "boosting", are almost entirely accounted for by the increase in fission energy, not the energy directly released by thermonuclear fusion.

How well boosting will work in a weapon of a particular design cannot be determined with high confidence by calculations, laboratory tests, or very low yield nuclear tests (i.e. at nuclear yields well below 1 kiloton) alone. Data from past nuclear tests incorporating boosting can help considerably in predicting the performance of new types of boosted weapons. Without prior experience of this kind, however, weapon developers would need to test boosted weapons to be confident of their performance.

Testing of boosted weapons requires high enough temperatures in the fissioning explosive to ignite the thermonuclear fuel. These temperatures depend on the unboosted efficiency of the weapon - the lower the efficiency, the lower the temperature in the fissioning material. For any particular type of design, there is some minimum threshold yield below which tests of boosting are not possible, because the temperatures are insufficient to ignite the thermonuclear fuel. This threshold yield decreases with increasing effectiveness of the fissile material assembly system in achieving high compressions, since higher compressions allow smaller yields to be achieved with the same efficiency.

The boosting threshold yield may be somewhat lower than 1 kiloton in some types of sophisticated weapons, but can be considerably higher in other types. In any case, deviations from idealized, calculated performance, along with the considerable complexity of the boosting process itself - involving important feedbacks between fission and fusion as the explosion proceeds - tend to make it necessary to test boosted weapons at full yield if they are to be put into weapon stockpiles. Such full yields are likely to be greater than 10 kilotons for many types of weapons.

9.3 Thermonuclear Weapons

These are defined here as explosives in which the energy produced by the thermonuclear reactions is a significant fraction of the total yield, as opposed to boosted weapons, where it is not. Thermonuclear weapons require a fission "trigger" to ignite the thermonuclear fuel. Vigorous efforts to develop pure fusion weapons that do not require fission triggers have so far apparently been unsuccessful.

It is difficult to imagine confident stockpiling of thermonuclear weapons by a country that has never tested any with yields substantially larger than the yields of their fission triggers. Furthermore, the ratio of the full design yield of thermonuclear weapons to the minimum, reduced test yield needed to provide high confidence of their performance, is smaller than for fission weapons. The reason is that thermonuclear weapons are substantially more complex and subject to performance uncertainties than fission weapons. Confident development of thermonuclear weapons is therefore likely to require nuclear tests with yields greater than a quarter or so of the full design yield. The minimum test yields would tend to be smaller for countries that have extensive experience testing thermonuclear weapons of various types than for countries that have not.

9.4 "Third Generation" Nuclear Weapons

Third generation weapons may be pure fission, boosted, or thermonuclear weapons designed to enhance the release of X-rays, neutrons, microwaves, unvaporized high velocity pellets, very high velocity plasma, or other forms of energy propagated in a particular direction from the explosion.[1] This is in contrast to most nuclear weapons that have been developed, for which the principal measure of performance has been the total energy released by the weapon when it explodes.

Full yield testing is likely to be required for all types of third generation nuclear weapons. One reason is that the partition of energy between different basic forms produced in a nuclear explosion is generally likely to be dependent on the yield. Another is that directional enhancement of a particular form of energy that is possible in one range of total yields may not be in a much higher or lower yield range.

Confident development of such weapons generally will also require many more nuclear tests than have been required for previous nuclear weapons. Complex measurements of the effects of specific types of energy, at various intensities, on various types of targets will be needed to assess the performance of the weapons. It can be argued that these effects cannot be confidently measured in underground tests, and therefore that confident development of third generation nuclear weapons will require resumption of nuclear testing in the atmosphere and in space.

9.5 Notes and References

1 For an overview on third-generation nuclear weapons, see: T. B. Taylor, Endless Generations of Nuclear Weapons, Bulletin of the Atomic Scientists, vol. 42, no. 9, pp. 12-15, Nov. 1986; T. B. Taylor, Third-Generation Nuclear Weapons, Scientific American, vol. 256, no. 4, pp. 22-31, April 1987

10. Recent Developments and Outlook for the Verification of a Nuclear Test Ban

Jeremy Leggett

10.1 Introduction

The USA and USSR agreed to re-open formal negotiations on the test ban on September 17, 1987, with the aim of first agreeing verification measures which would satisfy the Reagan Administration that ratification of the Threshold Test Ban Treaty (TTBT) and Peaceful Nuclear Explosions Treaty (PNET) was finally possible. (The TTBT was signed in 1974, the PNET in 1976). That done, the negotiations would proceed "step-by-step" to further intermediate limitations: in other words, to consider thresholds below the 150 kilotons set by the TTBT. The negotiations would have in mind the "eventual" goal of a Comprehensive Test Ban Treaty (CTBT), as part of "an effective disarmament process".

The bilateral negotiations began in Geneva on November 9th, against a backdrop of growing interest in nuclear testing limitations in the US Congress. Since 1986, efforts had been made in both House and Senate to cut off funds for tests above 1 kiloton, provided the Soviet Union did the same. (At the time negotiations were agreed, a Bill to this effect was going through the Senate). The first section of this review summarizes events relevant to test ban verification from the time that the first round of negotiations began.

The interest in Congress led to the publication on 26 May, 1988, of a report by the Office of Technology Assessment on the subject of "Seismic Verification of Nuclear Testing Treaties".[1] That report, which was the product of exhaustive hearings, constitutes a milestone in the long-running debate over test-ban verifiability. The second section of this paper summarizes the main conclusions of the OTA Report, and examines their implications.

In the final section, on prospects for test ban verification, the policy options are examined from the verification point-of-view, with emphasis on the relative merits of verifying a low-yield threshold (between 1 and 10 kt), and a CTBT.

Distinction has to be drawn between scientific opinions about the degrees of verifiability of the various options, and the political likelihood that they will be negotiated. This is so because of the current US Government belief that as long as a nuclear stockpile is needed it must be tested for effectiveness, reliability, survivability, and safety.[2] The controversy which that view has engendered is beyond the scope of this review. A good summary of the issues and arguments can be found in Fetter's book.[3]

10.2 Recent Developments

10.2.1 The Joint Verification Experiment

The Joint Verification Experiment (JVE) of August-September 1988, during which Soviet scientists measured a nuclear test on-site in Nevada and American scientists measured a Soviet test on-site in Kazakhstan, was proposed by the USSR to encourage the US Government to submit the TTBT and PNET for ratification. The US Government position is that hydrodynamic measurement (an on-site technique) is needed for measuring tests above 50 kt; the USSR argues that seismological measurement (a remote technique) is adequate for verification of the 150 kt limit of the TTBT, but has agreed in principle to the hydrodynamic measurement of tests.

On January 8-15, 1988, US Government scientists visited the Soviet test site at Semipalatinsk for the first time, to familiarize themselves with the operations there. Soviet scientists visited Nevada January 24-30, for the same reason, and for a demonstration of CORRTEX, the US hydrodynamic yield-measurement device. CORRTEX is an acronym for Continuous Reflectometry for Radius versus Time Experiments. It consists of one or more coaxial cables which are inserted into a satellite hole in the ground near (within metres) of the nuclear device, or in the burial shaft itself. The cables are attached to a battery-powered, suitcase-sized electronic unit, which remains above ground. Repetitive electrical pulses are sent down the cable at intervals of tens of millionths of a second. The underground explosion crushes the cable, producing a fast-moving short, which can be measured as a function of time to give a figure proportional (if the ground- and hole conditions are known) to the yield of the explosion. The device works well only above 50 kt, and operates best in the "strong shock" region closest to the explosion.[4] The Soviets use an instrument called MIS, which operates on the same principle as CORRTEX.

On April 14, seven Soviet officials visited the Nevada Test Site to observe the preparation of the satellite hole for the JVE. The Americans doing the same job in Semipalatinsk were dissatisfied with the Soviet drilling. In early May, they flew a complete drilling rig, and crew, into the Semipalatinsk test site, in order to drill the satellite hole themselves. The formal agreement to go ahead with the JVE was signed by Shultz and Schevardnadze at the Moscow summit on 31 May.

By the end of June, as the second round of nuclear testing talks came to an end, the two sides were near to agreeing a protocol for the verification of the PNET. Drafting on the TTBT verification protocol was to wait until after the JVE. They had exchanged data for five high-yield explosions per side: the historical yield values, geological conditions at the test site, and seismological measurements at five in-country stations. Such data, the exchange of which was formalized in the original TTBT protocol but never enacted because the treaty was not ratified, is valuable for the calibration of teleseismic measurements (measurements made outside the country, or at distances in excess of 2,000 km). The accuracy of the five historical values, of course, has to be taken on trust, so the exact role of the JVE remains somewhat unclear: a US official at the talks had earlier noted, for example, that a single explosion would not be enough to calibrate seismic equipment - six to twelve would be needed.[5] The values given by the Soviets for their 5 tests have been kept classified by the USA, though the senior Soviet negotiator, I. Palanykh, said late in 1988 that the Soviet side have no objection to the data being released.[6]

On 17 August, 1988, the US JVE explosion went ahead. Each side had a hydrodynamic cable in both the emplacement hole and the satellite hole, and also carried out teleseismic measurements at the five seismic stations for which historical data had been exchanged. The blast measured 5.4 on the Richter scale. An official said the bomb was around 145 kt, and US Government seismologists said that their measurements confirmed that the bomb was in the 140 kt range.[7] Hydrodynamic measurements were less accurate, according to information acquired by the New York Times: the CORRTEX cable in the actual explosion hole registered 163 kt, the cable in the satellite hole registered 155 kt, and the Soviet device apparently measured the blast at 170 kt.[8]

On 14 September, the Soviet JVE explosion went ahead. Yields were not divulged at the time, but a US Government official later said that the seismic estimates were close to the on-site measurements. The New York Times reported the on-site hydrodynamic measurements to be 115 kt (USA) and 122 kt (Soviet). The Government expert said that US seismic measurements put the test at 135 kt.[9] Within a few weeks, data from 118 seismograph stations had been compiled for the Soviet JVE. US seismologists calculated a body wave magnitude (mb) of 6.1. The yield deduced from both body waves and surface waves was 113 kt.[10]

In the few months before, more than 140 US technicians had lived and worked at the Soviet site, with only one problem: three officials were caught in July trying to smuggle illicit souvenir samples out in their luggage. The cost of the operation had been about $28 million, and will be about $10 million for every subsequent Soviet test for which the deployment of CORRTEX is deemed necessary.[11]

10.2.2 Controversy over CORRTEX

The basis for the Reagan Administration's insistence on using hydrodynamic measurements for verifying the TTBT and the PNET has been that CORRTEX is accurate to within a factor of 1.3, whereas seismological monitoring is only accurate to within a factor of 2. This is a view which is not shared universally. At the request of Congress, the Office of Technology Assessment conducted investigations into this and other aspects of test ban verification. Their hearings, conducted during 1987, were exhaustive - 50 seismologists, for example, offered opinions.[12] The report was released on 26 May, 1988. The Director concluded that the claim that teleseismic monitoring could only achieve a factor of two was "obsolete", because new methods could reduce uncertainties to 1.3-1.5. In other words, CORRTEX buys little if any advantage in yield estimation, a conclusion which the JVE figures quoted to the New York Times seem to bear out.

On October 6th, Milo Nordyke, a senior staff scientist and former chief of the Treaty Verification Program at the Lawrence Livermore National Laboratory, disputed that the CORRTEX method was needed for TTBT verification in testimony before the Senate Foreign Relations Committee. It was overly intrusive, he said, and should only be used in a limited number of experiments for the purposes of calibrating seismological measurements.[13]

In May, a leaked memorandum had shown that in 1982 the then Director of the Livermore Laboratory had questioned the need for CORRTEX even before the decision to seek improved verification for the TTBT had been taken by the Reagan Administration. The memorandum, said pro-test ban Congressman Edward Markey, showed that the Administration "has cynically used the CORRTEX proposal to stall progress towards any further limits on nuclear testing".[14] The President of the US Arms Control Association has described the use of CORRTEX in the JVEs as "a needlessly complex and circuitous assault on a largely manufactured problem".[15]

The JVE agreement stipulates that data from the JVE, and the previously-exchanged values for five tests per side, cannot be made public without the agreement of both sides. Currently, although the USSR has no objections, the US Government does. A law suit has been brought against the DOE under the Freedom of Information Act by the Natural Resources Defense Council (NRDC) and the Lawyers' Alliance for Nuclear Arms Control.[16]

The negotiations have been in abeyance during the transition period from the Reagan to Bush Administrations, with the final drafting of the TTBT and PNET protocols still awaiting completion. At the close of the third round of nuclear testing talks in December 1988, a senior White House Official was quoted in the Washington Post as saying: "The experiments essentially convinced each side that their own approach is best. We seem to be stuck."[17]

10.2.3 Non-Governmental Verification Developments

In parallel with the governmental negotiations about how to verify a 150 kt threshold, interest in test-ban verification by seismological means has been blossoming among non-governmental scientists, as the result of the success of a project initiated in 1986 by the Natural Resources Defense Council. The data being gathered as part of this effort are relevant to verification of a threshold much lower than 150 kt.

The NRDC, a private environmental group, signed an agreement with the Soviet Academy of Sciences for the setting up of three seismic stations near the Semipalatinsk and Nevada test sites. The stations were to be manned by academic seismologists from both the USA and the USSR. The three US stations were not allowed to operate when Soviet tests took place at the end of the moratorium, but at other times recorded valuable data, including incoming signals from US tests in Nevada. An experiment was staged in September 1987, wherein three chemical shots were let off on the test site, and recorded at the three sites, at an average distance of 190 km. The shots, consisting of two 10-ton charges and one 20-ton charge, were recorded clearly at all the stations. One arrived in the middle of signals from a large earthquake originating near New Zealand, and was still distinguishable.

A kindred experiment was mounted in Nevada in April 1988. Soviet seismologists set up their own equipment at Deep Springs, California, about 190 km west of the Nevada Test Site. On 29 April, two 10-ton explosions were set off near Reno, and on 30 April a 15-ton explosion. The scientists involved claimed that the Soviet seismic equipment had worked well, and that the experiment had confirmed - as expected - that in Nevada the Earth's crust absorbs seismic energy to a greater extent than it does around Semipalatinsk.

During August, five new NRDC stations were installed in the USSR, at Obninsk, Garm, Kislovodsk, Arti, and Irkutsk: all more than 1,000 km from Semipalatinsk. These stations were to monitor all Soviet tests from the JVE onwards. Like the three predecessors, all the five stations operate high-quality seismographs in boreholes, so cutting down the influence of noise on recordings. British university seismologists were also invited to install a seismic station at Garm, and did so during the summer.[18]

The non-governmental efforts had two principal effects. First, a large body of data is being collected from within the USSR, where once there was none. This will translate in time into quantification of some of the seismological arguments described below. Second, a large body of seismologists is being drawn into the subject of test-ban verification. US seismologists involved in the NRDC effort represent institutions, such as the universities of California, Colorado, and Columbia, and British seismologists from universities such as Glasgow, Leeds, and Leicester are now also working on data relevant to test-ban verifiability. In the early 1980s, research on test-ban verification outside the weapons laboratories was very rare.

10.2.4 The Ad-Hoc Group of Seismic Experts at the UN Conference on Disarmament

Multilateral efforts to encourage a CTBT have been stagnant for many years at the Conference on Disarmament (CD), despite the enthusiasm of many states for such an agreement. Where verification is concerned, however, one important body is contributing. The Ad Hoc Group of Seismic Experts, comprising seismologists from 25 countries, conducted a trial exchange of data using a global grid of seismic monitoring stations, between 15 October and 15 December 1984. A second such technical test (GSETT-2) is planned, with the transmission of more-extensive data, including waveforms (so-called Level 2 data), to Experimental Data Centres, and the transmission of results back to participating countries. A full-scale experiment is planned for 1990.[19]

10.2.5 Expenditure on Test-Ban Verification Research

In November 1987, the House of Representatives Select Committee on Intelligence published a report on the status of US infrastructure for arms control verification.[20] Among the conclusions on the test ban were:

> "U.S. seismic systems for monitoring nuclear testing were found to lag considerably behind the state of the art of seismology..." "U.S. efforts to address this shortfall were found to be inadequate..." "There are acute shortages of analysts trained in critical skill areas..." "Upgrades of these (seismic) systems have progressed slowly to date and have been characterised by mismanagement and contractual problems..." "Critical shortages of analyst personnel trained in seismology and geophysics exist within U.S. intelligence...".

The committee concluded that one agency should be set up, and that several hundred million dollars should be diverted to test ban research. At that time, some $70 million was directed to test-ban verification research annually, of which up to $40 million was dedicated to seismology. Subsequently, the situation has improved slightly: the DOD FY 89 Authorization Bill authorized $37.6 million to DARPA for improvements to nuclear monitoring capabilities (as well as requiring a Nuclear Test Ban Readiness Program of the Department of Energy).

Verification budgets for the USSR and UK are not available.

10.3 The Office of Technology Assessment Report and its Implications

The importance of the OTA Report, published on 26 May, 1988, springs from the exhaustiveness of the investigations which led to its production, and the fact that it tries to identify a consensus view among the US technical community, confining its conclusions to those which would be shared by the great majority of US experts.

The main conclusions, apart from the assessment of CORRTEX already discussed, can be summarized as follows:

1. There is no evidence that the USSR has cheated on the 150 kt limit of the TTBT, as the Reagan Administration has persistently alleged.
2. Cavity decoupling (exploding a bomb in a cavern deep underground so as to dampen the magnitude of the seismic waves produced) poses the only significant route to successfully violating a LYTTBT or CTBT, and this task could not be accomplished easily or cheaply.
3. It is not possible in practice to decouple a test above 10 kt, and then (between 2 kt and 10 kt) only in a salt dome.
4. A network of a dozen or so arrays outside the USSR would have an identification threshold (i.e. level above which it becomes routinely possible to discriminate explosions from earthquakes) equivalent to about 1 kiloton exploded in hard rock, or about 70 kilotons in a cavern (seismic body wave magnitude, or mb, 4.0).
5. For arrays inside the USSR, there is agreement that tests equivalent to several hundred tons in hard rock, or about 18 kilotons in a cavern, could be discriminated (mb 3.5). The threshold for detection would be much lower, equivalent to tens of tons in hard rock, or around a kiloton decoupled (mb 2-2.5). Detection alone is inadequate, of course: the number of small earthquakes which could in principle be decoupled explosions would swamp the system with false alarms. The identification threshold is set so much higher by the OTA because of the number of false alarms that would arise from chemical explosions, which occur in large numbers at levels below about 200 tons.
6. As the OTA Report points out, some seismologists believe that monitoring high seismic frequencies will allow routine discrimination of nuclear explosions of 1 kt decoupled and tens of tons in hard rock. Since there is no consensus yet on the practicality of this technique, the OTA Report observes that: "All of the considerations so far have not made allowances for increased verification capability afforded by high-frequency recording. The recent NRDC recordings to very high frequencies at distances of 200-650 km from three chemical explosions with yields of 0.01-0.02 kt are very impressive in this regard..." and "...if the monitoring system has even a modest capability to record frequencies as high as 10-15 Hz, the effectiveness of the decoupling would be greatly reduced". The data now being gathered in the USSR will go some way towards resolving this issue.

Summarizing, the authors of the OTA Report felt able to offer the following general conclusion:

> "Most experts agree that a high-quality network of internal stations combined with stringent constraints, could monitor a threshold of around 5 kt. Experts further agree that below 1-2 kt, monitoring would become much more difficult because additional methods of evasion are possible. Explosions of 1 or 2 kt could be decoupled not only in salt but also in other media such as granite and alluvium. At present, there is not a consensus that an internal network would be capable of positively identifying with high confidence all such evasion attempts. If such a capability is possible, it will require demonstration through practical experience of low-yield monitoring within the

Soviet Union together with a high level of negotiated supplementary measures to limit certain evasion opportunities."

The report has three important implications:

1. A platform of broad consensus has been established on which policy-makers can base decisions, and against which the rapidly-accumulating in-country data from the Soviet Union can be gauged.
2. It is clear that the overall prognosis for verifiability of a test ban, even a Comprehensive test ban, is good. There are no reasons, to do with verifiability, why the threshold cannot be lowered well below the 150 kt limit set by the current TTBT: certainly to 20 kt, probably to 10, and possibly to 5 kt. Only below 5 kt is there yet to be an identifiable consensus that seismology can be relied on to verify a ban effectively, and as more empirical data come in from the Soviet Union, this position will require regular review.
3. Techniques other than seismology apply to test-ban verification, and the OTA Report highlights the need for careful consideration of the role they can play, in an adjunct role to seismological monitoring, and in-concert with a cooperative approach to confidence-building measures (see below).

10.4 Prospects for Test Ban Verification

The technical capabilities of the USA and USSR to verify the Threshold Test Ban Treaty and the Peaceful Nuclear Explosions Treaty, by teleseismic monitoring alone, or by a combination of seismology and hydrodynamic monitoring, are such that they can be ratified immediately. The hydrodynamic technique, which only works well above 50 kt, buys little additional accuracy - certainly not enough to justify the huge expense involved in on-site hydrodynamic measurement of tests.

A 10-20 kt Threshold Test Ban treaty could be verified effectively by seismological means alone, given the facility for in-country seismic monitoring. However, testing at 10-20 kt would still allow the development of fission triggers for new strategic warheads, and as such would add little to the existing 150 kt constraint.

The OTA concluded that there might be monitoring problems, arising from false alarms, in the verification of a Low Yield Threshold Test Ban Treaty (LYTTBT) with a 1-10 kt limitation, and a CTBT. The view taken was that "...The monitoring burden placed on the seismic network by various evasion scenarios can be greatly lessened if seismology gets some help. Specifically, the structure of any treaty or agreement should create a testing environment such that ... a combination of prohibitions, inspections, and seismic methods will provide the desired high levels of verification

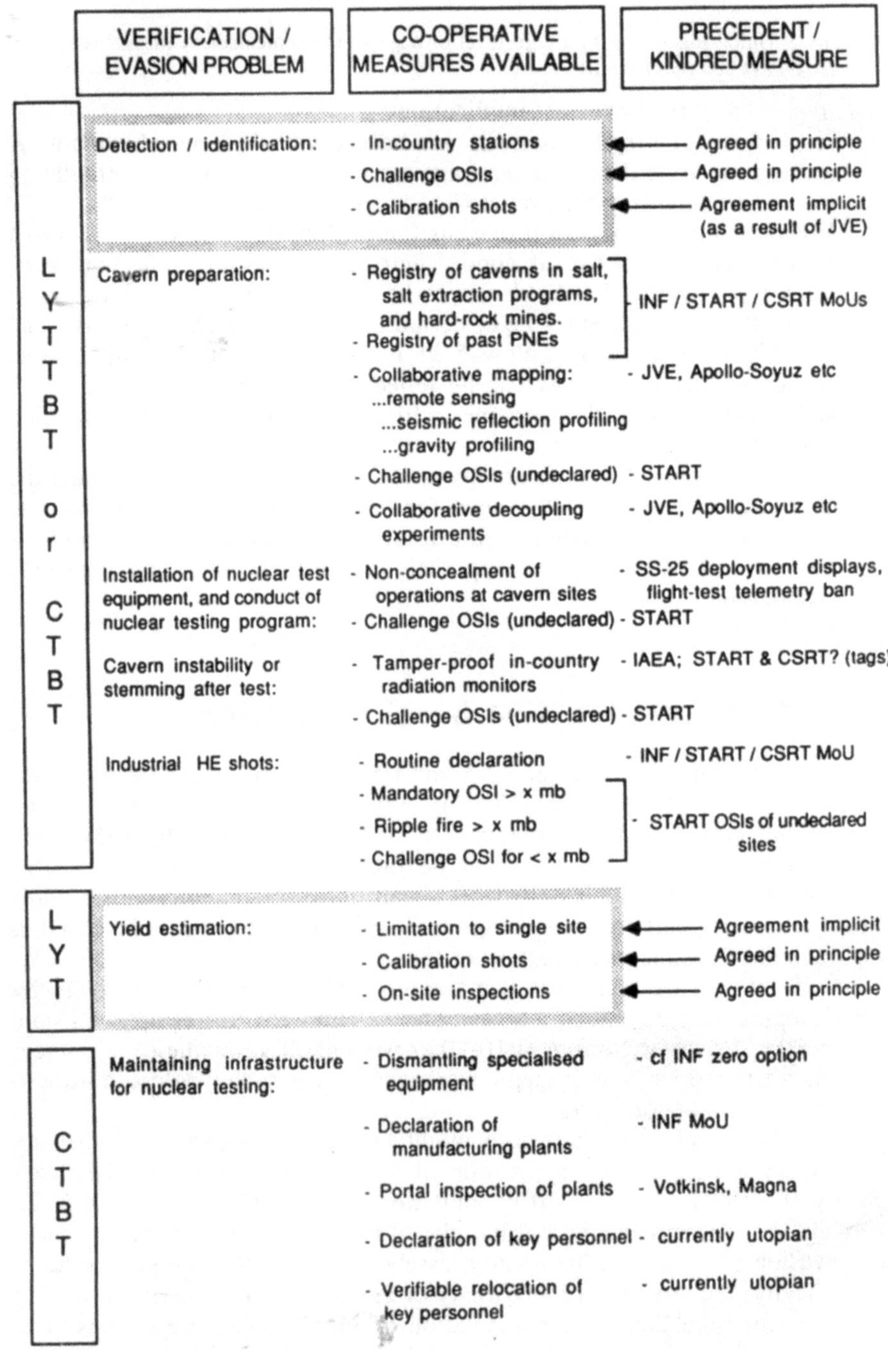

Fig. 10.1. Confidence-building measures which are available for the parties of a LYTTBT or CTBT (see text)

capability."

Verification has been defined as "...the establishment of truth or correctness of (something), by examination or demonstration."[21] This definition stresses the obligation of the potential evader to demonstrate compliance: after all, he presumably values the treaty and so has an interest in bolstering the confidence of the verifier. This can be done by means of confidence-building measures (CBMs). CBMs have been defined as "...those interstate national security undertakings designed or intended to increase the confidence in the 'good' intentions of potential adversaries.".[22] Figure 10.1 summarizes the CBMs which would be available to the parties to a LYTTBT or CTBT as a result of the geological and operational difficulties involved in staging a clandestine testing programme. They are discussed in detail elsewhere.[23]

Verification/evasion problems in the left column refer to difficulties both of effecting a successful clandestine programme (for the potential evader) and being absolutely confident that a clandestine programme is not taking place (for the verifier). It is the verifier's problem which must be accommodated in a successful treaty. The cooperative measures that are available in principle, which exploit the seismological, geological, and operational gauntlet which the evader would have to run, are listed in the middle column. A measure of their real-world feasibility is obtained by comparing them in "degree of intrusiveness" to analogous precedents and/or kindred measures agreed or under consideration in other arms control fora.

The most important potential CBMs are as follows:

Since the chief problem is cavity decoupling, the first option is to establish a register of caverns and mines. This would assist the task of photo-interpreters enormously. Fetter has argued that the job of checking for new engineering operations away from sites of normal human activity can be done using area-surveillance imagery from civilian satellites.[24] Close-look inspections, both of declared and undeclared sites, can be done with military satellites. A few challenge on-site inspections could be agreed to back up the surveillance process. Such OSIs would be capable of checking the nature of suspicious undeclared engineering operations over a salt dome, for example, or of ensuring that a site declared as a gas storage cavern was indeed that. This is not an overly intrusive arrangement when measured against those under consideration for the Strategic Arms Treaty. It involves, after all, mere holes in the ground. (Exclusion clauses could be negotiated to deter mischievous abuse: no caverns would be likely in the naval dockyards near Murmansk, etc).

A second obvious CBM concerns decoupling. The process has only been attempted once in the West, and then only at a level of 380 tons.[25] Can a cavern withstand a 10 kt blast in practice? If so, what decoupling factor does it achieve? How well do high frequency waves propagate from the explosions? To what distance can they be detected and identified? These questions, and others, can be answered by a relatively simple collaborative experimental programme.

Industrial explosions are the biggest single problem to be overcome. They can be declared as a matter of routine, with the option of an OSI for small blasts, and mandatory OSIs for blasts at levels equivalent in principle to decoupled nuclear bombs of worthwhile size. Additionally, they can be ripple fired: an aid to discriminating chemical from nuclear explosions.

On the basis of the "real world" analogues from other arms talks, only two of the listed CBMs can currently be regarded as impractical: the declaration and verifiable relocation (to other locations, and other jobs) of the (limited) number of personnel with nuclear-testing experience sufficient to attempt a clandestine programme. But this perception may change before too long: if states decide on an international treaty-regime which seriously comes to grips with the greenhouse effect, conceptions of national sovereignty will be forced to soften. The idea of checking the academic work-output of several dozen physics professors who once ran a nuclear testing programme will come to seem less utopian than it does today.

10.5 Summary and Conclusions

The JVE provided governmental verification experts with a means of comparing the accuracy of hydrodynamic and seismological techniques in verifying limits on testing near 150 kilotons. The results seem to favour the view expressed in a recent Office of Technology Assessment report on test ban verification, that seismological monitoring from afar can give just as good results as hydrodynamic monitoring on-site, at less cost and intrusiveness. Whatever the JVE has done for the science of verification, however, it has certainly provided the USA and USSR with scope for confidence-building.

The next step for the USA and USSR is to consider a threshold limitation below the 150 kilotons set by the TTBT. Recent contributions to the technical debate over verifiability centre on the relative merits of verifying various Low-Yield Threshold Test Ban (LYTTB) options in the range of 1-10 kt, or a Comprehensive Test Ban (CTB).[26] For LYTTB options, the principal verification advantage is the opportunity to minimize problems in discriminating earthquakes, chemical explosions and nuclear explosions. The principal disadvantages are the difficulty of accurate yield estimation at low yields, and permitting the maintenance of a testing infrastructure which could support a clandestine programme away from the agreed test site/s. For the CTB, the main advantage centres on the same "zero option" principle as applied for the INF Treaty: in other words, given the necessary cooperation, it is easier to verify total abolition of an activity than partial constraint.[27] This advantage, however, can only be realized if both sides take a positive approach to confidence-building measures. Precedents in the INF Treaty, and prospects in the strategic arms and conventional forces talks, suggest that such cooperation is possible in principle. Only the political will to "go zero" on testing is currently lacking. Perceived political expediency, rather than adequacy of verification, is likely to be the key constraint in the next stage of the test ban talks.

10.6 Notes and References

1 U.S. Congress, Office of Technology Assessment, Seismic Verification of Nuclear Testing Treaties, OTA-ISC-361, Washington DC: U.S. Government Printing Office, May 1988
2 Arms Control and Disarmament Agency, Arms Control Update, no. 8, August 1988
3 S. Fetter, Toward a Comprehensive Test Ban Treaty, Cambridge MA: Ballinger, 1988
4 F. K. Lamb, ACDIS WP-2-87-2, Urbana IL: University of Illinois Program in Arms Control, Disarmament and International Security, 1987
5 Interview, Arms Control Reporter, p. 605.B.76, 26 April 1988
6 New York Times, 11 September 1988
7 New York Times (note 6)
8 New York Times (note 6)
9 New York Times, 30 October 1988
10 L. R. Sykes, G. Ekstrom, Comparison of seismic and hydrodynamic yield determinations for the Soviet joint verification experiment of 1988, Proc. of the National Academy of Sciences of the USA, in press
11 M. Nordyke, Prepared Statement for the Senate Foreign Relations Committee, 6 October 1988; and: US Verification Method Disputed, Eos, American Geophysical Union, 8 November 1988
12 Eos (note 11)
13 Nordyke (note 11)
14 New York Times, 26 May 1988
15 S. M. Keeny, Notes from underground, Arms Control Today, vol. 18, no. 2, 1988
16 New York Times (note 6)
17 Washington Post, 16 December 1988
18 See: R. A. Clark, UK-USSR and US-USSR Joint Research Programmes in Seismic Verification, this volume, Ch. 11
19 UN Conference on Disarmament, Document CD/853
20 Report by the Permanent Select Committee on Intelligence, U.S. House of Representatives, 19 November 1987
21 F. R. Cleminson, E. Gilman, A conceptual working paper on arms control verification, Arms Control Verification Studies no. 1, Ottawa: Department of External Affairs, 30 pp., 1986
22 J. Macintosh, Confidence (and security) building measures in the arms control process: a Canadian perspective, Arms Control and Disarmament Studies no. 1, Department of External Affairs, Canada, 142 pp., 1985
23 J. K. Leggett, The role of confidence-building measures in verification of low-yield threshold and comprehensive test ban treaties, long discussion of this topic, prepared August 1988 for discussions at the Pugwash Annual Conference, Dagomys, USSR; and summary in: J. K. Leggett, Can we safely negotiate a treaty banning all nuclear tests?, The Lancet, pp. 674-5, September 17, 1988
24 Fetter (note 3)
25 The 1966 Sterling explosion, described in the OTA Report (note 1) and references therein

26 Phasing out nuclear weapon tests: A report to the President and Congress on Nuclear Weapons Policy, Special Publication, Washington DC: Natural Resources Defense Council, 1989

27 Leggett (note 23)

11. UK–USSR and US–USSR Joint Research Programmes in Seismic Verification

Roger A. Clark[1]

11.1 Introduction

Seismology - the study of the sources and propagation of elastic waves within the Earth - is the main means of monitoring compliance with arms control treaties concerning underground testing. Seismic waves can, depending on the explosive yield of the test and the conditions it is detonated under, be recorded at distances from a few kilometres up to thousands of kilometres from the test location. The present treaty regime, the 1974 Threshold Test Ban Treaty (TTBT) restricts weapons tests by its signatories (UK, USA, USSR) to an unlimited number of underground detonations each of no more than 150 kt (kiloton TNT-equivalent) yield. Comparable restrictions are placed on nuclear explosions for engineering purposes or Peaceful Nuclear Explosions (PNEs). The TTBT can be verified adequately using systems which do not intrude into the territory of the state being monitored (National Technical Means, NTMs) once data exchanges required in its Protocol have taken place. These will measure accurately the attenuation of seismic waves in the Earth below the US and Soviet test sites (since this biases the P-wave magnitude, from which yield estimates are made).

Proponents of arms control measures argue that limitations on testing are pivotal to meaningful moves towards nuclear arms reductions, and the USSR and USA agree in principle that they should move towards reductions in testing, and ultimately to a Comprehensive Test Ban (CTB); opinions differ, though, on exactly how to progress. The USA wishes first to ratify the TTBT (and has even argued that reduction in deployed weapons implies a need for maintaining or even intensifying its testing programme); others propose stepwise imposition of more stringent treaties, with lower yield thresholds and quotas on numbers and/or aggregate yields of tests. Treaties, such as a Low-Yield Threshold Treaty limiting explosions to a maximum yield of 10 kt, will require that a seismic monitoring network includes stations within the USSR, because the seismic signals from small explosions might be very weak and undetectable at more distant stations (especially in various suggested scenarios for attempted clandestine tests). This, then, demands that we know more of the geophysical characteristics of the USSR, because to design this network and assess its capability, we must know seismic background noise levels, wavespeeds and attenuation, structure of the Earth's crust and mantle, numbers of earthquakes and industrial explosions occuring, etc.

Suitable data for such assessments for the USA are in the public domain, but equivalent data have not, in the past, been released by the USSR through government or academic channels. Estimates of these parameters have nevertheless frequently been made, in order to predict the capabilities of various hypothetical networks, and it must be stressed that, until actual data are obtained, such predictions remain entirely speculative. Experiments at governmental level, via the Geneva Conference on Disarmament have been notoriously difficult to organize and infrequent. Only one has taken place, the 1984 "Level I Data Exchange", where arrival times and amplitudes of a few types of seismic waves were transmitted from and between a network of national and international data centres. The experiment was a qualified success, and an extension of this to transmit Level II data (full digital waveform streams, where the quantity of data will be over two orders of magnitude greater) should take place in 1989 or 1990.

11.2 USA-USSR Programmes

The Level I Data Exchange did not permit many of the issues in verification seismology to be resolved, addressing mainly the practicalities of large-scale near-real-time data transfer. This shortcoming has long been recognized, and in 1986, a US environmental pressure group, the Natural Resources Defense Council (NRDC), arranged directly with the Soviet Academy of Sciences to carry out a joint and reciprocal programme of seismic monitoring around each nation's main weapons test sites (Nevada & East Kazakhstan).[2] Three US-operated stations, within some 100-150 km of the East Kazakhstan test site, provided the first seismic data ever obtained from that region by Western seismologists. Because of its close proximity to the Soviet Test Site, though, the NRDC were required to halt recording during nuclear tests. While not prejudicing scientific aims of the experiment, this led to the USSR announcing tests in advance for the first time. A key element of the programme, both in the USA and the USSR, has been the detonation of "calibration shots" (chemical explosions of known yield) close to the Nevada and Kazakhstan Test Sites to quantify further the relation between explosion yield and wave amplitudes.

The NRDC-USSR programme has provided important data, for example on seismic noise levels and wave attenuation rates within that geological environment. It has also stimulated greater interest in verification research in the academic community - for example, useful techniques in location of seismic sources are being developed which incorporate particle motion and wave polarization data with the "traditional" arrival time information.

11.3 UK-USSR Programmes

From an early stage in the NRDC project, a UK involvement was sought and was agreed in principle with the Academy of Sciences in February 1987. After 18 months of fund-raising from non-government sources the British-USSR Seismic Verification Research Project (BSVRP) started operation in August 1988 at Garm, Tajikstan (Fig. 11.1) in time to record the 17 August Nevada "Joint Verification Experiment" (JVE) explosion (arranged by the USA and USSR to demonstrate on-site yield estimation techniques). The project is primarily University-based (at the Universities of Leeds, Leicester, and Glasgow) together with several eminent scientists now working independently. Its role at present is simply data acquisition - funding is insufficient to support an analysis effort - but this itself is significant political and technical progress. Data on magnetic tape will be archived at international centres in the UK, USA, and the Netherlands, as well as with Soviet colleagues. Garm is outside the military exclusion zone of 1,000 km around test sites, and so both Soviet and US nuclear tests and Soviet PNEs have been recorded - the first time such data have been acquired and released in the West.

One specific technical aim of the BSVRP is to examine the potential of "high-frequency" monitoring. The project is using a UK-designed and built Guralp Systems CMG-3V seismometer, typical of recent advances in low-noise instrumentation that allow recording of ground vibration and frequencies of up to 30 Hz and higher. The Guralp has a uniform magnification response between 0.03 Hz and 30 Hz. Paralleling this, new theoretical models of explosion and earthquake sources predict that explosion-generated seismic waves are richer in high frequencies compared to earthquakes and this difference should increase with increasing frequency. Thus, acquisition of high-frequency data will be a most promising development in monitoring explosions. In some geological environments, high-frequency waves have indeed been detected, contrary to traditional understanding of wave attenuation; studying them within all geological settings in the USSR is of great importance.

The Garm station, in the Pamir Mountains, is also in a region of current earthquake activity and of relatively soft sedimentary rocks. It is thus a candidate area for potential clandestine tests, detonated just as an earthquake occurs to "mask" the explosion seismic waves and/or "muffled" by detonation in cavities or soft rocks. It is vital, then, to document the earthquakes and industrial quarrying explosions that occur, and study the geological structures in that region. Detection capabilities at Garm could set the level of capability of a full USSR network.

Fig. 11.1. General location map of the USSR, showing the Test Sites on Novaya Zemlya and in East Kazakhstan. The Astrakhan/Azgir region of west Kazakhstan is the site of many engineering nuclear explosions (PNEs). The location of the UK station at Garm, Tajikstan, is highlighted

11.4 Preliminary Analyses of BSVRP Data

Data acquired by the BSVRP are currently being transcribed to computers in the UK. Some initial analyses have been made on nuclear explosions and a suite of 12 local and regional earthquakes.

Up to early December 1988, at least 9 nuclear explosions, from Nevada to Novaya Zemlya, East Kazakhstan, and engineering sites in Siberia have been recorded. Natural seismic activity in the region is high - at least 20-25 earthquakes/day in addition to less frequent industrial chemical explosions. Examples of a Kazakhstan nuclear test and of local earthquakes are shown in Figs. 11.2 and 11.3 respectively. Background seismic noise is dominated by 5-6 second period "ocean microseisms" of 150-200 nanometres peak-to-peak displacement. The RMS ground velocity of the noise, averaged over 64 second-long windows at 4 hour intervals, is typically 50-60 nanometres/second, rising to 80-90 nm/s in windy conditions. This is a fairly low figure, and P-wave arrivals of 1 nm amplitude have been easily detected. At frequencies (f) higher than the microseisms, noise levels reduce rapidly, as about f^{-3} up to some 10 Hz but with a more a complex frequency dependence thereafter.

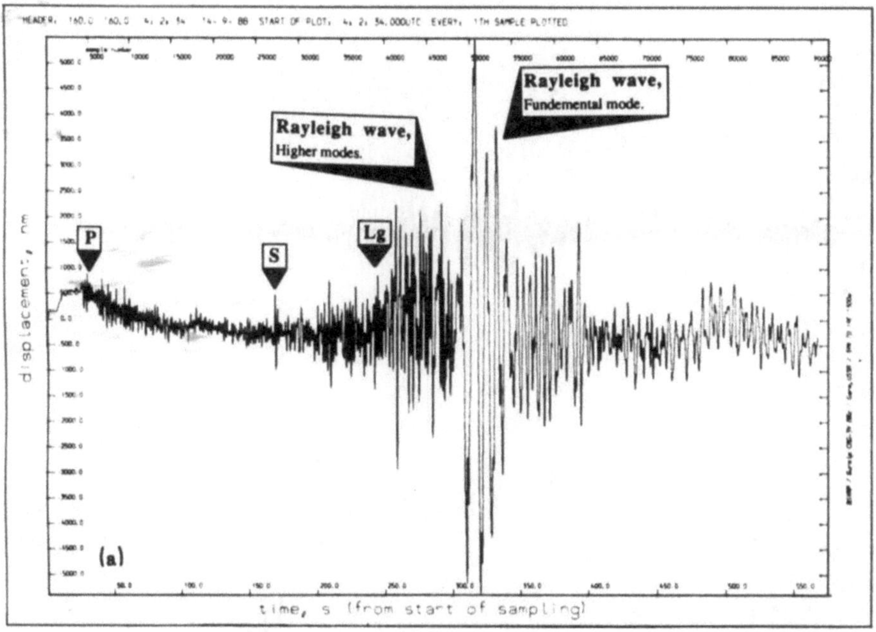

Fig. 11.2. (a) Displacement seismogram of the Kazakhstan JVE explosion of 14 September 1988, recorded by the BSVRP station at Garm (distance approx. 1390 km). Ground displacement is in nanometres. The low-frequency excursions near the start and end of the record are artefacts of processing and will be removed by further filtering etc. Various important seismic waves are indicated. Frequency spectra of the noise are compared to that of the P-wave (b), and Rayleigh (surface) wave (c) (next pages), both derived from the velocity seismogram. The P-wave (used to detect and locate events) has a signal-to-noise ratio of about 120; for the surface wave (the main tool for identification of explosions) it is about 40-50. The time windows on the seismogram used for spectral analyses are boxed. The magnification of the seismograph system varies with frequency, and has not been deconvolved from the record, but is uniform between 0.03 Hz and 30 Hz

These noise levels cannot yet be translated into a detection threshold, because although many explosions have been recorded, their yields are not publicly known. However, spectral analyses carried out so far show that for the 14 September 1988 Kazakhstan JVE nuclear explosion (Fig. 11.2a: believed to be some 100-150 kt yield), the P-wave was some 120 times larger than background noise (Fig. 11.2b). For the weak low-frequency surface waves generated by explosions, the signal-to-noise ratio (SNR) was about 40-50 (Fig. 11.2c). The USSR nuclear explosions, at ranges of between 1,400 km and 3,900 km, usually show optimum SNR at frequencies of some 1-10 Hz. Local earthquakes, at ranges of up to some 500 km or so contain signal energy above noise levels up to the 40 Hz limit of the recording system. Figure 11.3 shows examples of these. Similar processing of many more recordings will be required to separate the effects of source size,

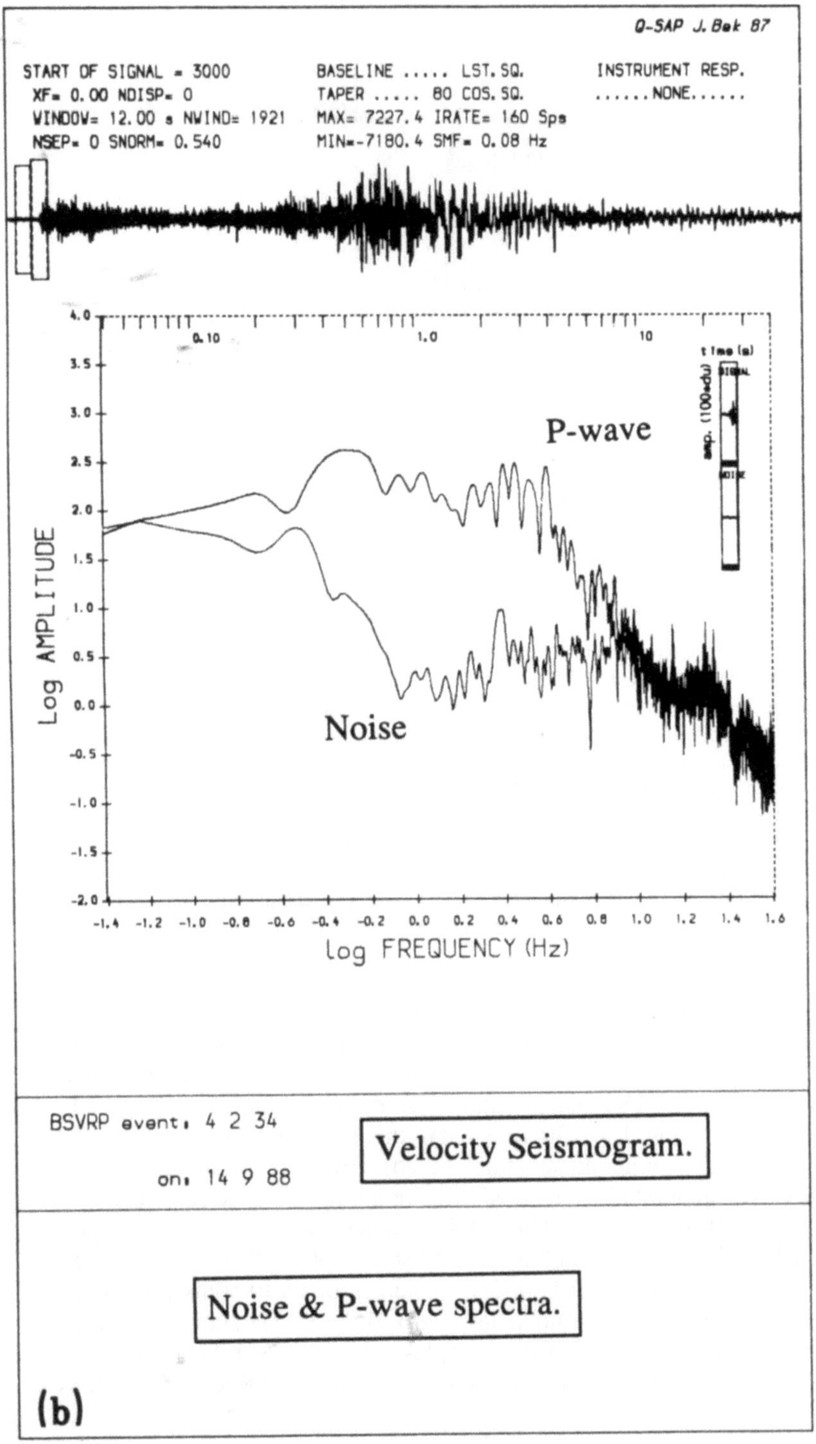
Q-SAP J. Bek 87
START OF SIGNAL = 3000
XF= 0.00 NDISP= 0
WINDOW= 12.00 s NWIND= 1921
NSEP= 0 SNORM= 0.540
BASELINE ..... LST. SQ.
TAPER ..... 80 COS. SQ.
MAX= 7227.4 IRATE= 160 Sps
MIN=-7180.4 SMF= 0.08 Hz
INSTRUMENT RESP.
......NONE......
P-wave
Noise
Log AMPLITUDE
Log FREQUENCY (Hz)
BSVRP event: 4 2 34
on: 14 9 88
Velocity Seismogram.
Noise & P-wave spectra.
(b)

(Fig. 11.2 b)

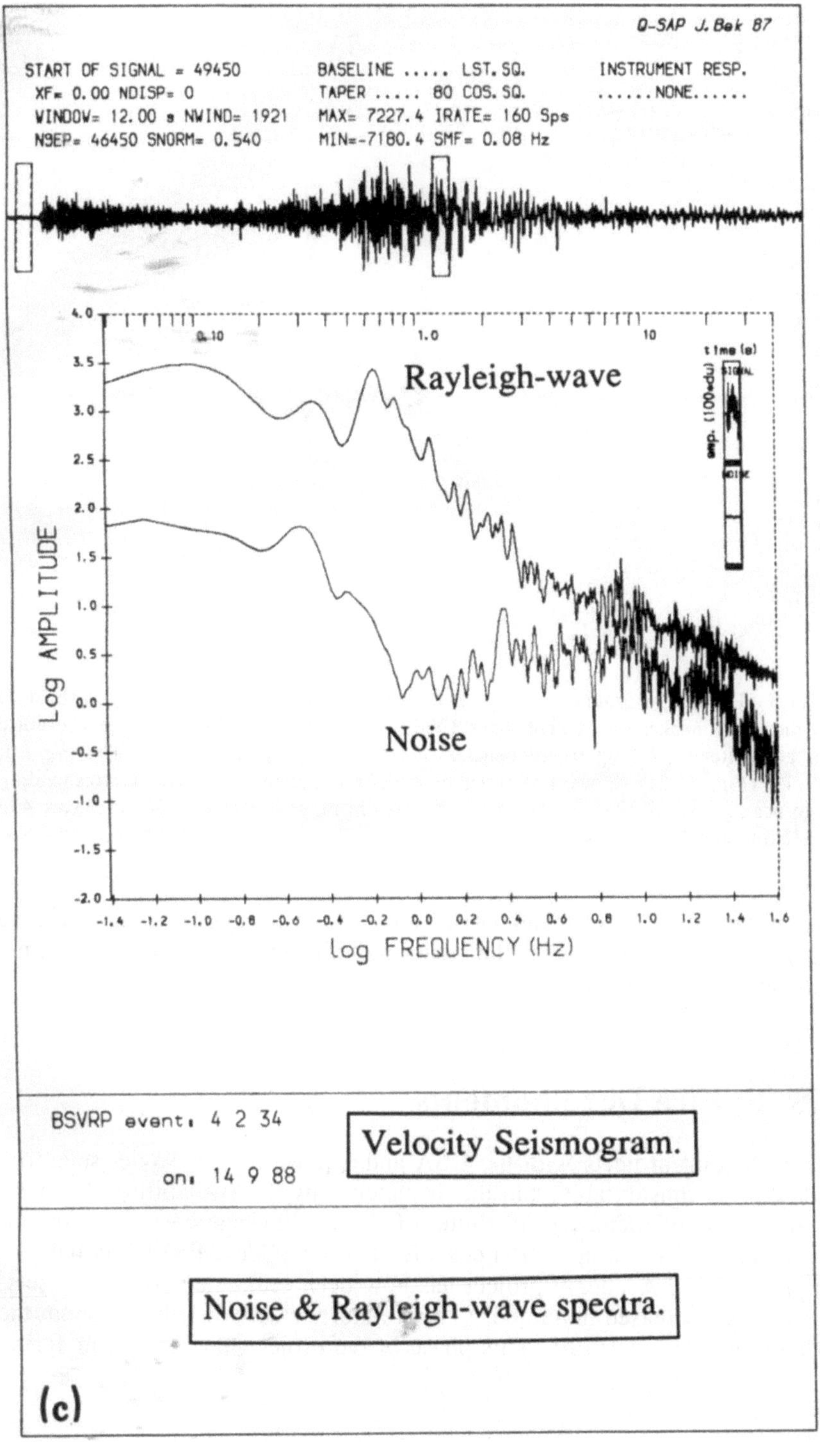
Q-SAP J. Bek 87
START OF SIGNAL = 49450
XF= 0.00 NDISP= 0
WINDOW= 12.00 s NWIND= 1921
N9EP= 46450 SNORM= 0.540
BASELINE ..... LST. SQ.
TAPER ..... 80 COS. SQ.
MAX= 7227.4 IRATE= 160 Sps
MIN=-7180.4 SMF= 0.08 Hz
INSTRUMENT RESP.
......NONE......
Rayleigh-wave
Noise
Log AMPLITUDE
log FREQUENCY (Hz)
BSVRP event: 4 2 34
on: 14 9 88
Velocity Seismogram.
Noise & Rayleigh-wave spectra.
(c)

(Fig. 11.2 c)

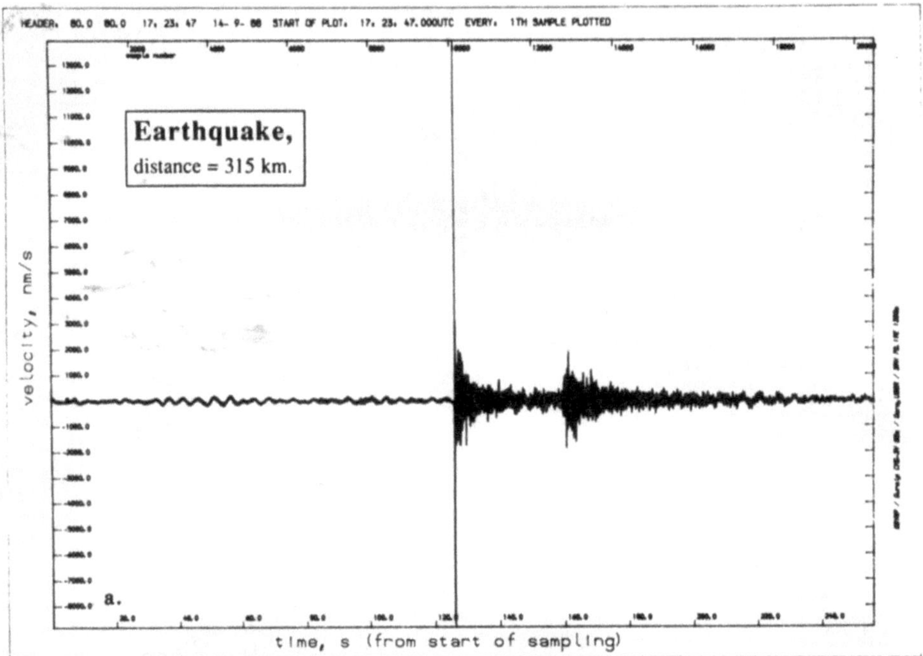

Fig. 11.3. Records of ground velocity (a, c, e) and associated frequency spectra (b, d, f) for 3 local earthquakes on 14 September 1988 (continued on next pages). Signal-to-noise ratios are better than 1 up to frequencies of 30 Hz and possibly higher. The event at 2219UTC (Fig. 11.3e) shows a remarkably impulsive and clear P-wave. Corresponding zero-to-peak ground displacements for the P-waves in Figures 11.3a, 11.3c, 11.3e are 409 nm, 336 nm, and 51 nm respectively

wave attenuation, background noise, and electronic system noise, and so enable us to understand what the treaty-monitoring capability of a seismic station at Garm would be.

11.5 Future Developments

Interest in joint projects with the USA and UK within the Soviet scientific community is considerable, and this is matched by the availability of funds - the Academy of Sciences Institute of Earth Physics cover all travel, subsistence, and site preparation costs for the NRDC and BSVRP collaboration. The NRDC-Academy project has now been redirected so that its seismic stations (increased in number to 5) are more widely distributed about the southern and central USSR. This phase of the project, due to end in spring

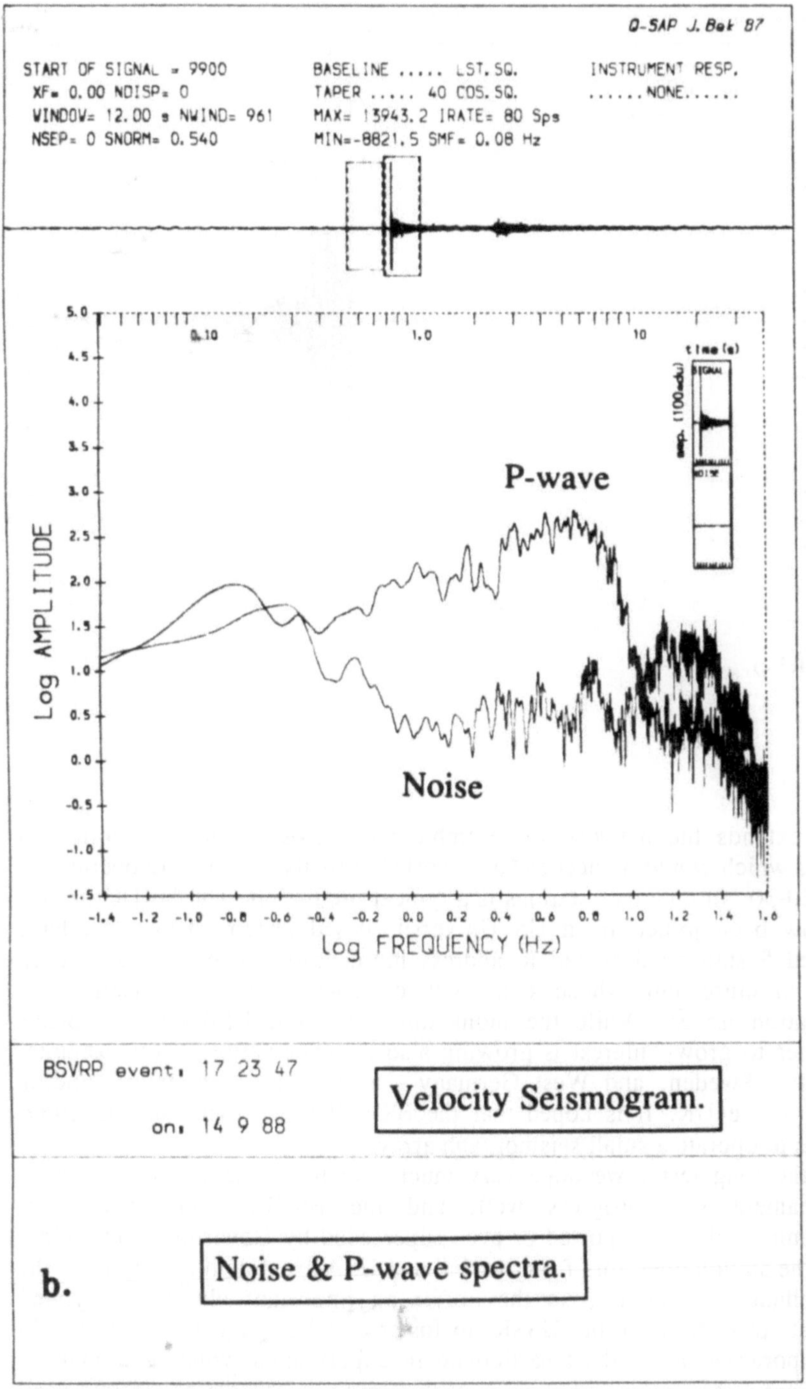
Q-SAP J. Bek 87
START OF SIGNAL = 9900
XF= 0.00 NDISP= 0
WINDOW= 12.00 s NWIND= 961
NSEP= 0 SNORM= 0.540
BASELINE ..... LST.SQ.
TAPER ..... 40 COS.SQ.
MAX= 13943.2 IRATE= 80 Sps
MIN=-8821.5 SMF= 0.08 Hz
INSTRUMENT RESP.
......NONE......
0.10
1.0
10
time (s)
P-wave
Noise
Log AMPLITUDE
log FREQUENCY (Hz)
BSVRP event: 17 23 47
on: 14 9 88
Velocity Seismogram.
b.
Noise & P-wave spectra.

(Fig. 11.3 b)

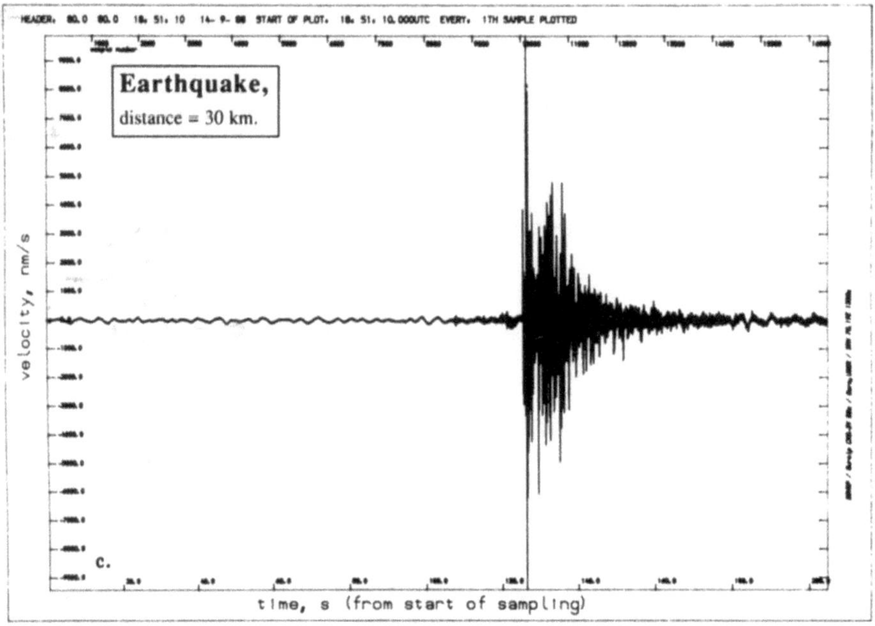

(Fig. 11.3 c)

1989, extends the network to resemble more closely the distribution of stations which would be needed for a workable treaty-monitoring operation - some 20-30 "in-country" stations is a typical prediction. The NRDC project has now been joined by a US University-based group, IRIS, who have installed 5 stations designed to address more academic problems of deep Earth structure but whose data will doubtless contribute usefully to verification issues. While the momentum for joint USA-USSR projects continues to grow, interest is growing also in other nations - for example, Australia, Sweden, and West Germany - as well as with major funding bodies in the UK. It is hoped that the BSVRP will expand during 1989, perhaps to operate a small seismograph array.

In the long term, we hope very much that the Geneva Conference on Disarmament will progress well, and the BSVRP and IRIS/NRDC programme will be supported or even superseded by Government participation. The station operators from the UK and USA are drawn mainly from the postgraduate community, so they offer an important chance for young scientists to experience the USSR, to foster collaboration with their Soviet contemporaries, as well as participate in experiments which will make a significant scientific and political contribution to the debate on seismic verification of nuclear testing treaties.

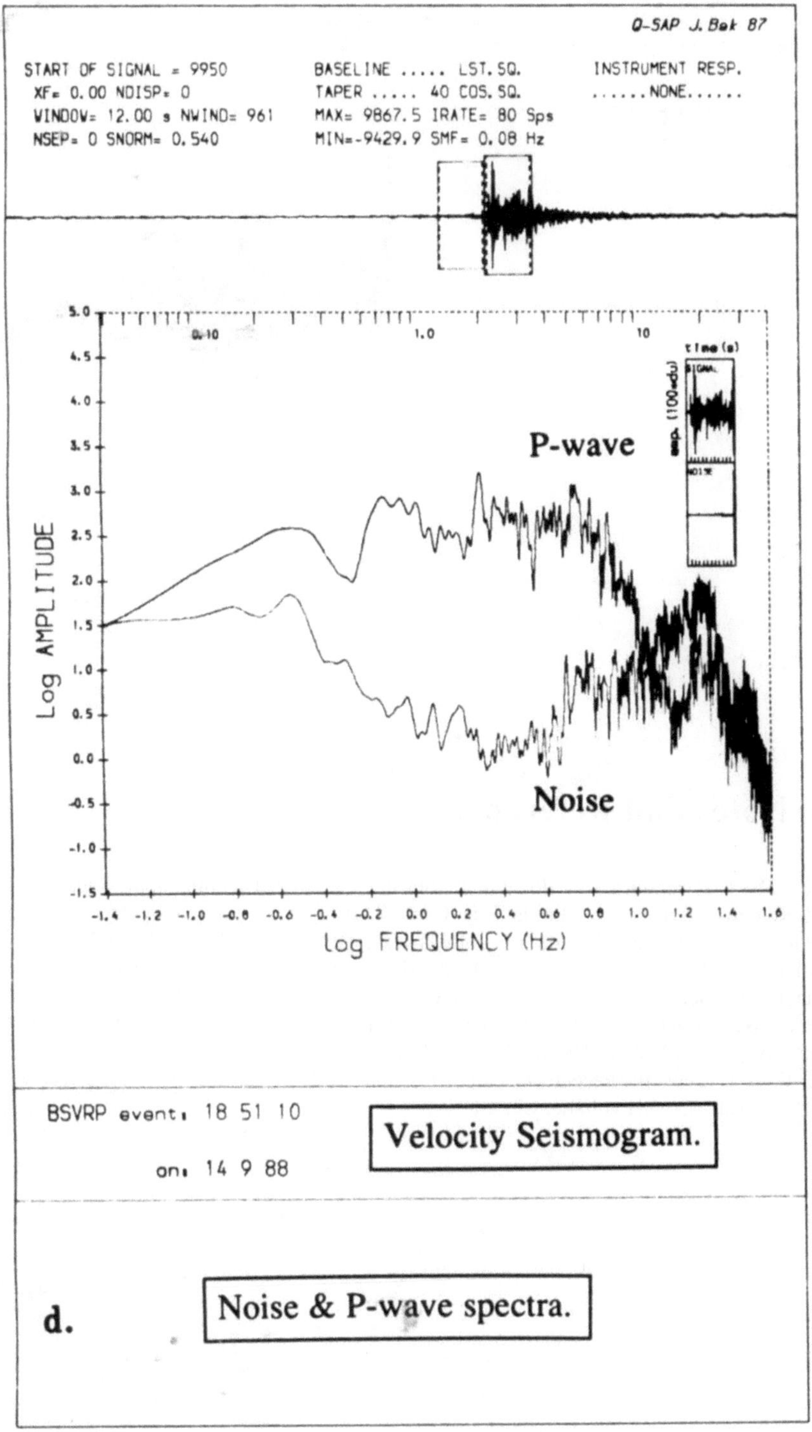
Q-SAP J. Bek 87
START OF SIGNAL = 9950
XF= 0.00 NDISP= 0
WINDOW= 12.00 s NWIND= 961
NSEP= 0 SNORM= 0.540
BASELINE ..... LST. SQ.
TAPER ..... 40 COS. SQ.
MAX= 9867.5 IRATE= 80 Sps
MIN=-9429.9 SMF= 0.08 Hz
INSTRUMENT RESP.
......NONE......
P-wave
Noise
Log AMPLITUDE
Log FREQUENCY (Hz)
BSVRP event: 18 51 10
on: 14 9 88
Velocity Seismogram.
d.
Noise & P-wave spectra.

(Fig. 11.3 d)

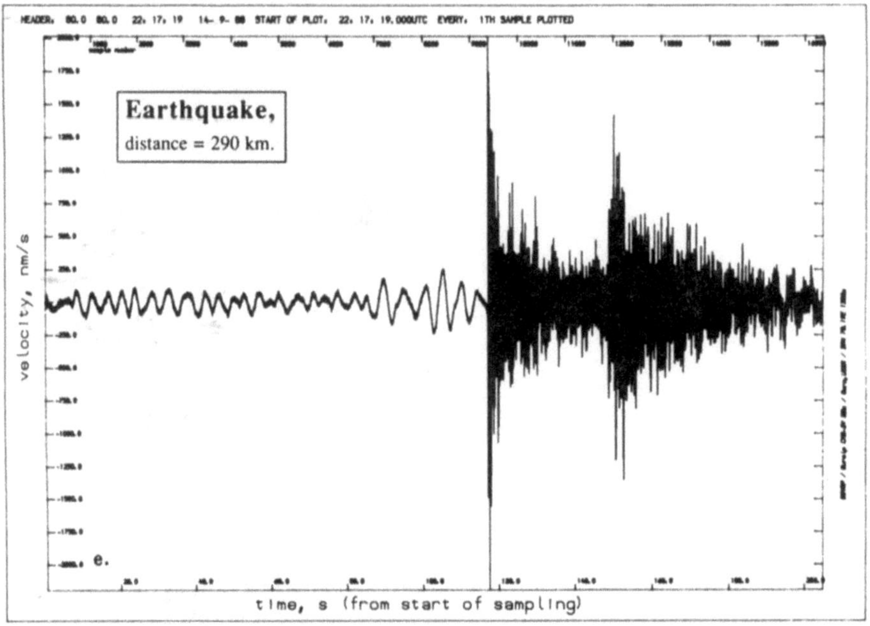

(Fig. 11.3 e)

11.6 Notes and References

1 The BSVRP is supported by donations from the Cadbury, Hickinbotham, Puckham, and Rowntree Trusts, and Leeds City Council. The project is led by Drs. David Davies and Frank Barnaby; its seismologists are Drs. Aftab Khan & Peter Maguire (University of Leicester), Dr. Ben Doody (University of Glasgow) and Mr. David Francis and the author at the University of Leeds. The station has been operated by Mark Askey, Janusz Bak, David Campden, Paul Coward, Kevin Davies, Fred Kirkwood, Angela McMahon, Duncan Moss and Peter Rands

2 See also: J. Leggett, Recent Developments and Outlook for the Verification of a Nuclear Test Ban, this volume, Ch. 10

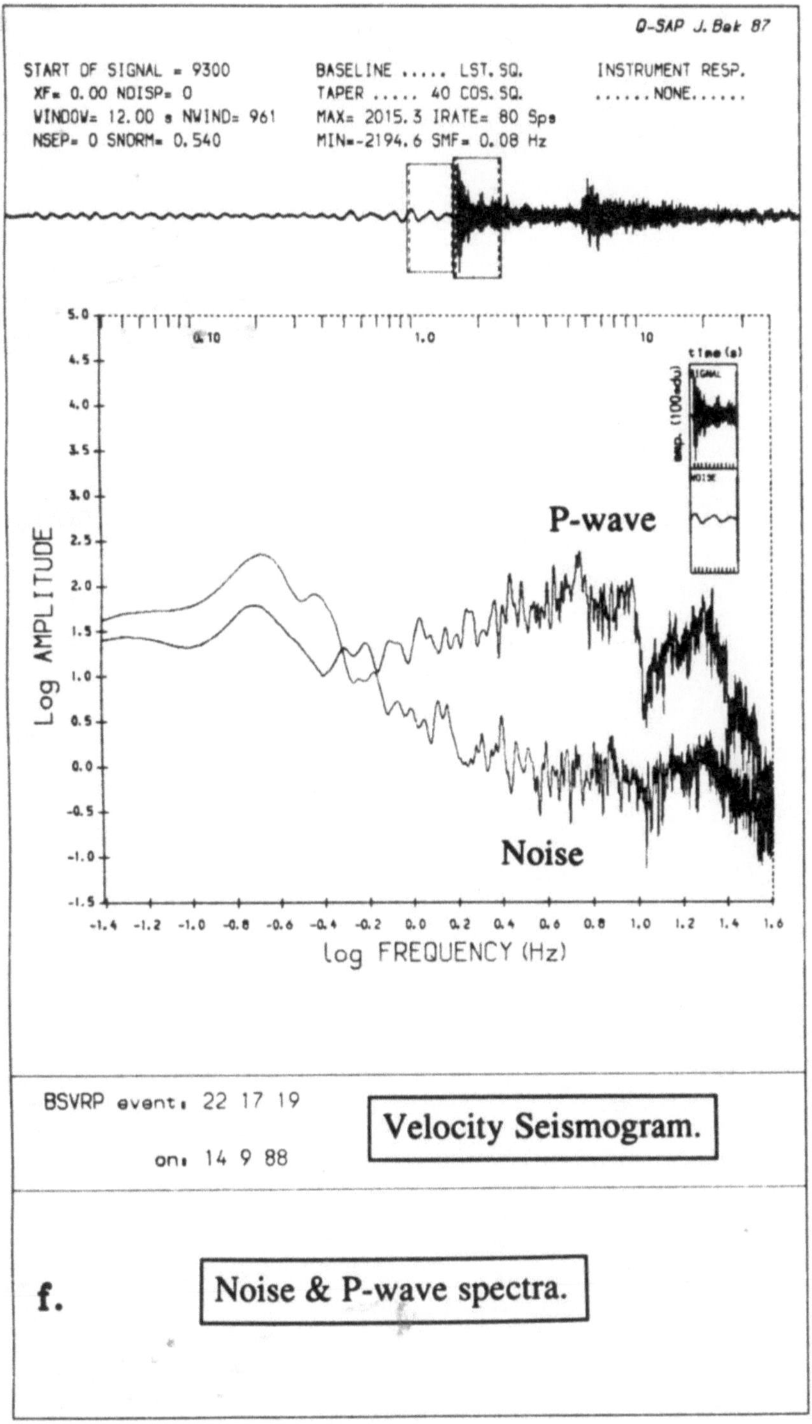
Q-SAP J. Bak 87
START OF SIGNAL = 9300
XF= 0.00 NDISP= 0
WINDOW= 12.00 s NWIND= 961
NSEP= 0 SNORM= 0.540
BASELINE ..... LST. SQ.
TAPER ..... 40 COS. SQ.
MAX= 2015.3 IRATE= 80 Sps
MIN=-2194.6 SMF= 0.08 Hz
INSTRUMENT RESP.
......NONE......
0.10
1.0
10
time (s)
SIGNAL
NOISE
amp. (100edu)
P-wave
Noise
Log AMPLITUDE
log FREQUENCY (Hz)
BSVRP event: 22 17 19
on: 14 9 88
Velocity Seismogram.
f.
Noise & P-wave spectra.

(Fig. 11.3 f)

Part IV

Control of Space Weapons

12. Recent Developments in Space Weapons

John Pike

12.1 Introduction

The past year has brought forth a considerable clarification and elaboration of the future course of the Strategic Defense Initiative in the United States, particularly the two review processes that were completed this year.

The first was through the Defense Acquisition Board, which is the primary council at the Defense Department that makes decisions and recommendations about military hardware development and procurement efforts. The first review by the Defense Acquisition Board on the Strategic Defense Initiative was held in the fall of 1987, and it approved an initial architecture for the Strategic Defense Initiative program. The principal problem that was encountered in the Defense Acquisition Board review 1987 was the very large cost associated with the Phase One deployment programme. The original cost estimate for deploying the Phase 1 of the Strategic Defense System was somewhere in the vicinity of approximately $120 to 150 billion. Nearly everyone associated with the process came dowm with an acute case of sticker shock, and the general reaction within the Pentagon was that there just is not enough money to go ahead with this programme.

At that time General Abrahamson, the director of the programme, proposed initial deployment of six elements for the first phase of the Strategic Defense System (see Table 12.1).

1. The Boost Surveillance and Tracking System (BSTS), a network of up to a dozen large sensor satellites located in geosynchronous orbital altitudes, with responsibility for initial detection of missile launches and hand-off of initial target tracks to other systems.
2. A mid-course Space Surveillance and Tracking System (SSTS), a constellation of several dozen satellites located in altitudes of several thousand kilometres that would track post-boost vehicles, reentry vehicles and decoys.
3. A Ground-based Surveillance and Tracking System (GSTS) that would perform the same mission as SSTS, using a probe that would be fired onto a ballistic trajectory by a large sounding rocket.
4. The first layer of defence of weapon systems would be the Space Based Interceptor (SBI), as many as 3,000 kinetic kill interceptor rockets fired from 300 satellite Carrier Vehicles in low Earth orbit.

Table 12.1. SDI Phase 1 System Concepts

System Element	Functions
Boost Surveillance and Tracking System (BSTS)	Detection of missile launches Acquisition and tracking of boosters and PBVs Kill Assessment
Space-Based Surveillance and Tracking System (SSTS)	Acquire and track PBVs, RVs, and ASATs Discrimination
Ground-Based Surveillance and Tracking System (GSTS)	Acquisition Tracking Discrimination
Space-Based Interceptor (SBI)	Disabling of Boosters, PBVs, RVs, and ASATs Sensors on carrier vehicle (CV) could provide enhanced midcourse sensor capability
Exoatmospheric Reentry Vehicle Interceptor System (ERIS)	Disabling of RVs in late midcourse
Battle Management/Command and Control, and Communications (BM/C3)	Man-in-the-loop control Engagement management Maintaining track data Target assignment Communications

5. The second layer of interceptors would be the Exoatmospheric Reentry Vehicle Interception System (ERIS), several thousand long-range ground-based kinetic kill interceptors.
6. The entire system being coordinated with a ground and air-based battle-management system.

12.2 Recent Changes in the SDI Programme

There have been a number of changes in the fine grain detail of this architecture over the last year, all ultimately driven by the desire to get the cost down to a level that is not obviously implausible.

The Boost Surveillance and Tracking System and the Space Surveillance and Tracking System satellites have been reduced in capability. The SSTS satellites in particular will be significantly smaller than was originally planned, with only half the initially contemplated mirror aperture. Greater reliance is going to be placed on the Ground-based Surveillance and Tracking System to perform the mid-course discrimination function. And a new element, a Ground Based Radar, an X-band system formerly known as the Terminal Imaging Radar, has also been introduced for both late mid-course and terminal phase tracking and discrimination.

The most notable in the recent changes in the architecture for the initial deployment of the Strategic Defense System is a major reduction in the number of planned Space Based Interceptors. The original plan called for approximately 3,000 interceptors to be carried on approximately 300 satellites. The current plan calls for somewhat less than 1,500 interceptors deployed on 147 satellites, each with 10 interceptors. The number of Exoatmospheric Reentry Vehicle Interception System rockets has also been increased by 70% over the number that was originally planned a year ago.

The result of all these changes is that the SDI Organization currently estimates that deployment of this initial phase of the Strategic Defense System, rather than requiring $120 or 150 billion, could be accomplished for slightly less than $70 billion. One of the reasons for this 50 percent cost reduction is the reduction in the number of some of the more expensive elements of the system, such as the Space Based Interceptors. The cost of individual elements has been reduced by reducing the size and complexity of the Space Surveillance and Tracking System.

Some elements have been dropped completely. Under the 1987 plan an entirely new heavy-lift launch vehicle, developed under the Advanced Launch System programme, would have been required to place the Space Based Interceptors into orbit. Now that the number of Space Based Interceptors has been significantly reduced, there is no longer a requirement for a new heavy-lift launch vehicle for the initial deployment of the Phase 1 system.

These changes have also been accompanied, perhaps not surprisingly, by a certain stretching out of the schedule for the programme. Under current plans a final go-ahead for deployment of the system would not be given until 1994, and the system itself would not achieve an initial operational capability until probably around 1998 with full operational capability not coming until early in the 21st century.

12.3 A New Political Strategy for Star Wars

The other main milestone, and significant change in the programme over the last year, was the report of the Everett Panel of the Defense Science Board, which had been charged by the Defense Acquisition Board in 1987 with reviewing the status and prospects of the Strategic Defense programme. This Defense Science Board panel released its report in the spring of 1988, and the report quickly gained considerable political currency.

The Everett report marks a fundamental departure in the game plan for how one would go about making the decision, or, in this case more appropriately, the decisions, for deploying a Strategic Defense System. Since President Reagan announced the Strategic Defense Initiative in 1983 the game plan for the decision on deploying a Strategic Defense System remained constant. Reagan, in his March 1983 speech announcing the SDI, said that he wanted a research programme that would enable "a future Congress and a future President" to make the decision for deploying the system.

For the last five years the subtext of the SDI debate was that at some undetermined point in the early 1990's there would be a single grand decision in which the United States would commit to deploy a large Strategic Defense System, going from no defences to large defences, and going from having an ABM Treaty to not having an ABM Treaty. This is the type of "Early Deployment" decision that former Secretary of Defense Weinberger advocated with singular lack of success.

It is becoming increasingly apparent that the American political decision making process is extremely adverse to making such very large, very momentous decisions. To the extent that the decision to deploy Strategic Defenses is cast in such terms, that decision is probably just too big of a decision to expect the American political system to make. The strategy outlined in the Everett Panel was to break the decision up into a series of small decisions, each of which would be not unpalatable, in and of itself. The Panel concluded that, although the final goal of the process would be a decision to deploy a full phase 1 Strategic Defense System, in the meantime, there were a series of smaller decisions that might profitably be made.

One of the Panel's recommendations was deployment of the Boost Surveillance and Tracking System (one of the sensor elements of the Phase One Strategic Defense System) to provide improved early warning of missile attack, and enhanced intelligence collection and verification capabilities. Deployment of the ERIS exoatmospheric reentry vehicle interception system for the anti-satellite mission was another recommendation, as well as deployment of small numbers of several of these systems as part of what Senator Nunn has suggested might be an Accidental Launch Protection System.

This strategy of several incremental decisions leading to the ultimate goal of extensive strategic defences raises several concerns. Much of the press reporting around the Everett Panel took as its central thrust the proposition that Star Wars is dead, since the rhetoric is no longer focused on Reagan's initial vision of "rendering nuclear weapons impotent and obsolete." But the notion that, since the rhetoric is no longer that of the magic astrodome, strategic defence technology is no longer a matter of serious concern, is fundamentally incorrect. Precisely the opposite is true. This new political strategy for deploying strategic defenses is probably even more worrisome than the old strategy. The magic astrodome rhetoric was so obviously absurd that it was fairly easy to point out its absurdity. On the other hand, some of the proposals of the Everett Panel, particularly the call for deploying improved early warning satellites, have a surface logic that will prove much more difficult to challenge. It is quite possible that over the next several years the Congress will decide affirmatively to begin deploying some of these systems, particularly the Boost Surveillance and Tracking System and the ERIS for the anti-satellite role.

12.4 Star Wars and the ABM Treaty

It is apparent that the debate over the broad reinterpretation of the ABM Treaty has been successfully concluded. The Congress has made it very clear that it regards the reinterpretation of the ABM Treaty as being without historical or legal merit. The Congress will not approve funds for SDI tests that are rationalized as being consistent only with a broad interpretation. Unfortunately, the Congress has made it equally clear that as long as the Administration is prepared to claim that a test under the SDI program is consistent with the traditional interpretation of the ABM Treaty, the Congress is unlikely to interfere in that testing process.

There are a series of tests that one would anticipate seeing over the next several years that would raise questions about compliance with the ABM Treaty. The most immediate area of concern is the initiation of field testing the Airborne Optical Adjunct (AOA) programme in 1989. The Airborne Optical Adjunct is not the first air-based sensor aircraft that the United States has tested (see Fig. 1). The TRIM U-2 flew over 25 years ago. NASA's Kuiper Astronomical Observatory has been in operation for a number of years, and the Optical Airborne Measurements Program aircraft began operations in 1988. But AOA is certainly the largest and the most capable of these aircraft, and as it begins its testing programme in the summer or early fall of 1989, it will raise questions about whether this air-based sensor system is capable of substituting for an ABM radar, which would be inconsistent with Articles V and VI of the ABM Treaty.

Thus far at least the Congress has been extremely reluctant to take affirmative action on this programme, and in all probability the testing programme is going to continue. Some of the other near-term tests that are a source of concern about potential compliance appear to be receding into the future. The Laser Integrated Space Experiment, a space-based chemical laser experiment originally planned for orbital testing in 1992, has been delayed by at least a year. The initial flight test of the Boost Surveillance and Tracking System, which would have also raised concerns about testing space based components, also appears to have been delayed.

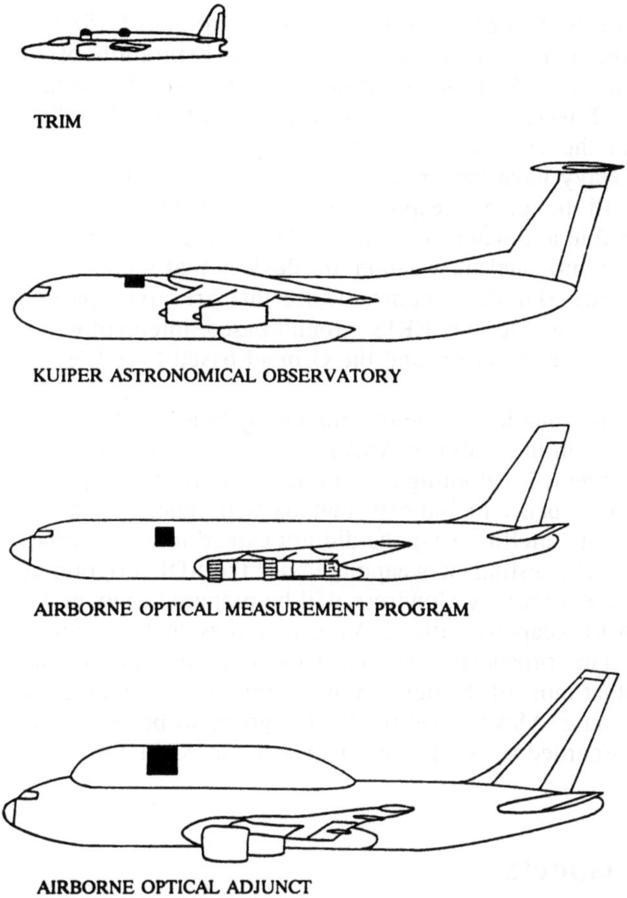

Fig. 12.1. US aircraft carrying optical sensors

12.5 Anti-Satellite Weapons

However, an area of equal arms control concern, anti-satellite weapons, seems to be an area of growing interest on the part of the American military. Over the last several years the anti-satellite situation has been fairly stable. The Soviet Union initiated a moratorium on the testing of anti-satellite weapons in 1983 that the United States Congress subsequently reciprocated.

Two events in 1988 cast into doubt the future in the area of anti-satellite weapons. In 1988 the Congress failed to re-extend the ASAT testing moratorium. There was an attempt to make it a permanent moratorium rather than one that would have to be renewed every year. This attempt failed, and in the process the annual moratorium was not renewed for 1989. This has not caused as much concern as it might, because at the beginning of 1988 the Air Force cancelled the F-15 ASAT programme. The expiration of the ASAT testing moratorium in 1989 probably is not going to result in the initiation of American ASAT tests, simply because there aren't any ASATs to test. But in the wake of the cancellation of the F-15 ASAT programme the Air Force, Army and Navy have engaged in a lively competition to determine which service will be given responsibility for developing a more advanced and more capable anti-satellite system. The leading candidates for a new and more capable anti-satellite weapon are devices that are currently being developed under the SDI Programme. Those that are given particularly prominent mention include the ERIS ground-based interceptor, the MIRACL ground-based chemical laser, and the Ground-based Free Electron Laser programmes.

The prospects for a continuation of the informal bargaining mode of arms control that has succeeded in the realm of ASATs over the last several years are not bright. The Congress, in adopting a moratorium on the testing of the F-15 system, was at great pains to indicate that its willingness to impose limits on the testing of anti-satellite weapons did not extend to a willingness to impose limitations on the testing of weapons under the SDI programme. Thus, there is little prospect that the Congress will be prepared to extend an ASAT test moratorium to weapons with ASAT capabilities that are part of the SDI programme. The prospects for that happening are particularly remote, given the fact that most of the devices that would be tested under the SDI programme which have ASAT capabilities, are going to be tested in a mode that is generally regarded as being consistent with the ABM Treaty.

12.6 Future Prospects

The prospects for space weapons arms control over the next several years are mixed. The bad news is that the Airborne Optical Adjunct is going to start testing next year, a potential violation of the ABM Treaty. The good news is that some of the other tests that would raise compliance questions have been delayed. But the really bad news is that the existing, relatively stable, tacit arms control regime on ASATs is likely to erode over time. A variety of new anti-satellite weapons are in the works and at this point it is politically unlikely that this new round of ASAT weapons will face the same sort of congressional restraints that were imposed on the F-15 ASAT system.

13. Verification of a Ban on Space Weapons

Stanislav N. Rodionov

13.1 Introduction

Although the title of this paper implies that a kind of international agreement on space weaponry ban is considered as possible, I am not going to go into details of this hypothetical treaty. Nor will I give any formal definition of space weapons; instead, I shall try to present some general (from the physical point of view) considerations of the possibility to verify a ban on a specific class of space weapons, namely those that are based on "new physical principles". But first I want to consider three possible methods of verification: ground to space inspection; space to space inspection; and on-ground (pre-launch) inspection.

The first method was described in detail by Roald Sagdeev at the Hamburg meeting two years ago.[1] Its capabilities are limited, of course. But, on the other hand, this method is relatively non-intrusive and it does not need any political agreement to be used.

The second method seems to be rather "exotic". Technically it is very sophisticated but its capabilities are limited mainly by the so-called "interactive discrimination" approach (identification of nuclear warheads on board the spacecraft of the other side). It should be noted that only a small part of all spacecraft might be inspected by this method.

Pre-launch inspection, on the contrary, is of a global nature. Each launching site location is well known to the international community, therefore, in the frame of the corresponding treaty, each launch may be potentially inspected with a preconditioned degree of intrusiveness.

13.2 Components to be Verified

Among the components of space weapons which should be verified, one may imagine the following set: platform, kill mechanism, energy system, "SATKA" system, C^3I system.

How may each of these components be verified at pre-launch inspection?

1. A platform should have to capability of carrying large masses of any material (and correspondingly large volumes), of high manoeuvrability, of fast retargeting, and so on. Therefore, these platforms should have large dimensions and have additional technical features typical of high-thrust engines. They would thus seem to be easily detectable. The

inspecting side can ask for an explanation why such a highly dedicated system is included into the payload, and if the explanation were unsatisfactory the inspecting side may insist on a delay or even abolition of the launch. But one should have in mind that such a platform may be constructed in space (for example, within a manned space station) using smaller elements which were not suspected by pre-launch inspectors.

2. Kill mechanism systems are supposed to be easily inspected on the basis of some specific feature. For example, for lasers large powerful mirrors are typical. For neutral particle beams rather long linear accelerators are neccessary. Electromagnetic guns should also have large dimensions. The only case when a quick look does not give any effect are normal rockets-interceptors. If one has in mind a final homing mass of, say, 10 kg, and final speed of a few kilometres per second, then corresponding rockets would be of a rather small scale and might be easily hidden. It seems now that the best way to verify deployment of small rockets in a huge payload would be to use X-ray transmission techniques in a way that would not be intrusive for the inspected side.
3. Practically for all "exotic" weapons power and energy consumptions are extremely high, which gives a real chance for verification. Our estimates showed that each shot of a laser or particle weapon needs about 100 megajoules of energy, corresponding to several hundreds of kilogrammes of the best chemical fuel. Of course, the system of energy supply should have several specific elements which are relatively easily observable:
   - energy source (MHD, nuclear reactor, combustion chamber, chemical reaction chamber and so on ...);
   - thermal energy/electricity conversion unit - its efficiency would not be very high, therefore, large cooling surfaces may be needed (a real hint for inspectors!);
   - electrical energy storage units (batteries, capacitor banks, induction coils and so on). These units are supposed to be rather massive, and it is not so easy to hide them from inspection. The batteries are "time-inertial" and cannot be used for producing high-power levels. For example, perspective lithium batteries promise to increase the energy capacity up to 10 MJ/kg, but the discharge speed is too low: it corresponds to a power capacity of only 1 kW/kg.

     Capacitors have a relatively small energy/weight ratio - up to 10 kJ/kg. This value may be increased by one order of magnitude in inductive coils. Rotating wheels are reported to have an even higher ultimate energy/weight ratio. All these devices have not only large sizes but other specific design features, so their visual identification does not seem to be a very serious problem.
4. The SATKA (Surveillance, Acquisition, Tracking, and Kill Assessment) system is a vital part of any weaponry unit. It consists mainly of any kind of radars, active and passive laser locators and rangemeters, high-speed computers, and so on. In principle, radar systems have relatively

large mirrors and dishes which may be considered as a specific signature for some SATKA devices. The real problem is that the kill mechanism system and the SATKA system may be separated spatially (deployed on different spacecrafts). This option makes the whole verification process more difficult but not impossible.

5. The C^3I (Command, Control, Communication, Intelligence) system is important for the large weapon constellation but not for a single weapon device. It consists mainly of software and, therefore, is very unsuitable for verification unless the inspection would be very intrusive.

13.3 Parameters

The main parameters which determine the military potential of any space weapon are:

1. For laser weapons it is the brightness, determined by: energy delivered in a single shot; mirror size; operational wavelength.
2. For particle beam weapons: energy delivered in a single shot; accelerator energy (i.e. length); accelerator emittance angle.
3. For railguns: energy delivered in a single shot; gun length; projectile mass.
4. For rockets: projectile mass; rocket size (before launch).

As shown before, linear dimensions may be controlled even from long distances.

There are also some "hidden" parameters: laser wavelength, accelerator emittance angle, mass of projectiles. One of the ways to resolve this problem may be that a corresponding political agreement would oblige the inspected side to provide information about these parameters. In other cases, an inspecting side might evaluate corresponding potentials using extreme values for particular parameters. For example, one might assume a wavelength of 0.3 μm for the laser, zero emittance angle for the accelerator and so on. These assumptions would give an overestimation. If the inspected side would not agree with these evaluations, it must present the proof of real performance for all doubtful devices.

13.4 Ban on Nuclear Power in Orbit

There is one action that may have a significant impact on the space weapon ban. On May 13, 1988 the Federation of American Scientists and the Committee of Soviet Scientists Against the Nuclear Threat held a press conference at the National Press Club at Washington, D.C. to announce the joint proposal for a ban of nuclear power in earth orbit. The proposal was relatively simple: to ban nuclear power in orbit while preserving the option of its use for civilian scientific purposes, such as deep space missions. This would serve the dual purposes of protecting the earth environment from radioactive contamination in case of an accident and providing a restraint on the militarization of space. Two recent disasters - Challenger and Chernobyl - forced us to rethink the use of nuclear power in space before there is a quantum leap in risk. Such a ban would be readily verifiable because of the strong infrared, gamma and neutron radiation emitted. We have even experimental confirmation of radiation detection from the nuclear powered satellite Cosmos 1900.

It is very interesting to note that a new physical mechanism was discovered which increases the background for scientific observation in the X-ray and soft gamma-ray regions of the electromagnetic spectrum. Gamma-rays from operating reactors have an average energy of about 1 megaelectron-volt. This means that there are a lot of gamma-rays whose energy exceeds 1 MeV. These gamma-rays are capable of producing electron-positron pairs in the reactor body. Escaping positrons move along magnetic lines in the earth magnetosphere and drift across these lines. So the positron cloud following the reactor is spread on rather long distances exceeding the range of "direct visibility" of the reactor by corresponding detectors. If a scientific satellite would cross the positron cloud not too late after its formation, the positrons coming onto the detectors and the satellite body would produce the so-called "annihilation gamma-line" with an energy of 511 keV. Many scientists have been looking for this line for years in the hope to prove the presence of antimatter in our universe. Nuclear reactors in space spoil this kind of scientific activity, producing artificial background events similar to the expected ones.

13.5 Notes and References

1 R. Sagdeev, Verification of a Space Weapons Ban, in: H. Spitzer (ed.), Scientific Aspects of the Verification of Arms Control Treaties, Hamburger Beiträge zur Friedensforschung und Sicherheitspolitik, nos. 19 and 20, Hamburg: Institut für Friedensforschung und Sicherheitspolitik an der Universität Hamburg, June 1987

Part V

Conventional Stability in Europe

14. Concepts of Conventional Stability and Reductions of Arms in Europe

Alexei Arbatov

14.1 Principles of Conventional Stability

I would like to begin with some general principles of stability which can be applied to the conventional arms reduction.

Notions of parity and stability were not inherent in the conventional arms balance, they were imported from strategic arms control. And when we are trying to apply those notions to the conventional arms balance, we meet a lot of difficulties, because strategic nuclear weapons are quite different from conventional weapons, both in their capabilities and the general principles of their use, or threat of use of these weapons.

On the other hand, there are certain similarities. And since we are trying to apply general principles of stability to the conventional arms balance, we should probably first seek the similarities between the strategic arms balance and the conventional arms balance. There is no need to describe all the differences and asymmetries, both technical and geographic, which exist between NATO and Warsaw Pact armed forces. But it is clear that neither NATO nor the Warsaw Pact has any plausible political intention to attack each other. This political stability, already existing in Europe, is the necessary precondition for conventional arms reductions. Otherwise all talks would be doomed to failure and it would not be worth starting them. On the other hand, the absence of aggressive intentions in both alliances is not in itself a guarantee that there can be no military confrontation in Europe. The problem is that both NATO and the Warsaw Pact have within their armed forces very powerful conventional weapons, very vivid offensive elements and offensive strategic concepts. Contemporary conventional armed forces have high a degree of mobility and can be used very quickly, especially in the forward area with their combat-ready forces.

By analogy with the strategic balance we can say that the sources of instability in Europe lie in the possible inadvertent escalation of a crisis. A situation in which in a political tense situation both sides will find themselves before a very dramatic and dangerous dilemma: either wait for the other side to act and risk catastrophic defeat, or preempt the other side. This is the dilemma which exists in the strategic offensive balance, and it exists also in the conventional balance. But the problem with conventional weapons is that many actions needed to bring them up to higher states of alert, to move them forward for defence, and to mobilize them, are indistinguishable from a defence point of view. Both sides will suspect the worst possibilities, and both sides will interpret the actions of the other, even though purely defensive, as preparation for offensive operations.

14.2 Objects and Difficulties of Negotiations

The increasing destructive power of contemporary weapons, their great range, and their high mobility, make the situation less stable. This is the main object to which negotiations should address themselves. Reduction of numbers, and limitations on deployment are not the goal itself, they are tools to achieve the main goal, which is to eliminate those capabilities and those elements of the military balance which may produce inadvertent escalation in a crisis, and bring both sides to war because of political misperception and fear that waiting might mean catastrophic defeat. Only on this basis, having this main goal in mind, all other issues may be addressed, including reduction of resources which are associated with conventional forces, increase of mutual confidence, and so on. Having said this about the similarities between conventional and strategic nuclear balances, it is important to note that many other elements are very different. Those elements predetermine the necessity of quite different approaches. On the technical level the difficulty lies in the fact that there are no offensive or defensive conventional weapons. All conventional weapons are multi-purpose. Their character depends not on their technical characteristics, but on their deployment, their cooperation with other elements of armed forces, their training, degree of readiness, proximity to the forward area, and so on. In this respect, on the technical level, the offensive or defensive character of a gun depends on whether you stand behind the gun or in front of it.

From this point of view the negotiators will face a very difficult task, because by dealing with instruments of negotiations, which most probably will primarily be numbers, they will have to address questions which are largely non-quantifiable. They will have to address offensive and surprise attack capabilities which depend not so much on numbers of weapons, but on other factors, such as deployment, possibility of reinforcement with rear echelons, cooperation between different armed forces, rear infrastructure, military art and many other factors. This is the first difficulty.

The second difficulty which has to be taken into account exists not on the technical, but on the tactical and operational levels. With the present fire power and mobility of armed forces, it is very difficult to imagine a defence of the type of World War I, static defence. Both alliances rely in their defensive planning on mobile defence. And mobile defence has as its priority such military actions as meeting an engagement, flanking attacks, and counter-attacks against penetrations. This means that on the operational level there would be weapons, forces, and plans, which are indistinguishable between offensive and defensive characteristics.

The third level of difficulty is a strategic one. The basic question is: if you have managed to counter-attack and get the aggressor out of your country, what do you do next? Do you stop at the frontier and wait until the other side recruits its forces and attacks again, maybe much more successfully this time, or do you follow on on the territory of the other side and finish the war by destroying the enemy? The history of the two world wars, and of many local wars, provides evidence that it is impossible to stop at the border, that the only way to ensure that the aggressor does not attack again is to follow on. On the other hand, to have the capability not only to repel the attack and to get the aggressor out of your territory, but to follow on, means quite a substantial military potential with very strong offensive elements. This would naturally be perceived by the other side as a potential of unprovoked surprise attack.

Thus, the negotiators are faced with a series of very difficult questions, dealing not only with technical problems to which we are paying justifiably a lot of attention now, but with many principal questions. Technicalities, such as what to do with reduced equipment, are secondary; they will have to be addressed, but only after having agreed on some principal points. In this respect there are several general principles which might be a basis for the stated goal of the Vienna mandate, i.e. to reduce conventional arms and armed forces in order to eliminate capabilities for surprise attack and large-scale offensive operations.

First of all, although all weapons are multi-purpose in their character - even landmines can be used offensively, for instance, to cover the flanks of penetrating forces -, certain categories of weapons when mass-deployed forward in superior numbers, are much more suitable for offensive than for defensive operation. The difference is not either-or; it is rather a difference of degree. But nevertheless it is a very important difference. Such weapons, as tanks, large calibre artillery, combat helicopters, surface-to-surface missiles, strike tactical aircraft, when massed in superior numbers within reach of the forward area, present a much more offensive than a defensive threat.

Second, carefully carved out sub-zones of reductions may disengage forces from the forward area, which means that the forces will have a much smaller capability of quickly engaging each other. This is important because surprise attack is greatly dependent on the capability to concentrate forces in the forward area and to attack on unexpected sectors of the front.

Third, a reduction in offensive forward deployed elements will by definition decrease the necessity of a strong counter-offensive capability, which may be perceived by the other side as an offensive capability. Forces for counter-attack will be retained, but they will be much smaller in size, because they will have to deal with smaller offensive threats, and they will be less perceived by the other side as a threat to it.

Fourth, reserve forces necessary for counter-offensive may be deployed very deep in the rear. Their use for a surprise attack would require much time, giving a clear signal to the other side, and time to prepare, thus denying any advantage to the aggressor.

14.3 Stability-Oriented Reductions of Armed Forces in Europe

I would like now to suggest a purely illustrative variant of how the conventional arms reduction agreement might look. I want to emphasize that nobody is trying now to forecast how negotiations will go, or to influence negotiations; it is just an illustration of an analytical approach to this very difficult question.

When we are defining stability or instability of the strategic offensive balance, we have rather well worked out modelling which includes different coefficients and definitions, distinguishing between forces which are stabilizing, i.e. which are not very good for surprise attack, and those which are very threatening, presenting a threat of a disarming first strike.

It may be wise to apply the same methodology to conventional forces, though of course in reality this will be very different. There are four destabilizing parameters which may classify conventional forces according to the degree of their danger, as instruments of possible surprise attack, as distinguished from forces which are more suitable for defence. The first, is the character and numbers of weapons when massed and concentrated. Second is proximity to the forward area. Third is the degree of mobility. And fourth is the degree of combat readiness. Those four points, if given reasonable substance, could create a formula which, however conjectural, may be applied at least to analyze the characteristics of this or that balance. By definition, agreements should address those four points and deal with them in such a way that the offensive quality of those four points will be reduced; this will produce a cumulative, synergistic effect in reducing the general offensive capabilities.

If we base our approach on those principles we could imagine, for instance, a zone of the greatest military concentrations, which certainly is in Central Europe. We call it the Jaruzelski zone; it includes the NATO guideline area, which was discussed in the Vienna negotiations in the 70s, as well as Denmark and Hungary. In this Jaruzelski zone approximately 50% of the tank assets and 40% of the heavy artillery of NATO and the Warsaw Pact are concentrated, and here the deepest reductions should be applied. As deep reductions I suggest 50% of combat-ready forces of both sides, taking as a starting point the level of the respective inferior side. This would leave both alliances approximately 5,000 tanks and 2,000 heavy artillery pieces each in this Jaruzelski zone in combat-ready forces. We can call this the "zone of prevention of surprise attack", according to the formulation of the Vienna mandate. The kind of reduction mentioned above would greatly reduce offensive surprise attack capabilities in this area.

The second zone is broader, and may be called the reinforcement area. It may include, for instance, the territories of Great Britain and France in the West, and in the East the territories of the three western military districts in the Soviet Union. This is purely illustrative; other areas can be chosen, for instance, it is possible to include also Italy, and the Kiev military district of the Soviet Union. This will be the subject for negotiations. Not many forces in combat-ready formations are concentrated in this area, approximately 15 to 20% of tanks and artillery, but these can be brought very quickly into the forward area. Those weapons will therefore also have to be addressed on the same basis: reduction to equal ceilings in combat-ready formations. Since this zone is further to the rear, reductions may be smaller, 30% for instance.

The third area includes all the other parts of Europe from the Atlantic to the Urals; this can be called the zone of large-scale offensive operations. In this area great assets are deployed, both of tanks and artillery, and other offensive systems, but they would require quite a long period of mobilization, because most of the units are not kept in a state of very high readiness. Therefore, in this zone the reductions can be even smaller, for example 20%, on condition that units with weapons in reserve, or low state of readiness, are not brought to a higher state of readiness. This is a very important point, remembering the fourth point of instability, namely that both sides have verifiable insurance that reserve forces are not brought to alert status.

15. Enhancing Conventional Stability in Europe

Heinrich Siegmann[1]

15.1 Introduction

Militarily speaking, conventional stability (CS) in Europe would be achieved if each side possessed military forces capable of defending their territory without posing a credible threat to the other side. In other words, the defensive capabilities of both sides would have to exceed the offensive capabilities of their antagonists even under worst-case assumptions. From the NATO viewpoint, its member countries must feel secure, especially against the possibilities of surprise attack and large-scale offensive operations stemming from the WTO's numerically superior conventional forces. From the WTO perspective, the territories of its member countries must be protected, even against merely limited objectives of the opponent, as well as in the case of adverse domestic conditions.

Let me make some general comments about this theoretical regime of defence dominance. Clearly, stability must not be equated with balance or parity. Perfectly balanced forces might be highly unstable, such as in a duel situation. We should also note that even numerically inferior forces have proven capable of very successful attack such as the German tank forces invading France and the Soviet Union in World War II. Thus, reductions alone cannot be the objective of arms control talks.

A second comment concerns the distinction between "offensive" and "defensive" weapons. One can argue that any weapon system can be used both offensively and defensively. There is simply no purely defensive weapon. In World War II the Germans had to rely heavily on their interceptors - generally considered defensive - for escorting their offensive bombers in the Battle of Britain. Similarly, Rommel's forces employed anti-tank weaponry very effectively during their forward thrust in Northern Africa. Still, it is fair to say that certain weapons can be applied with greater or lesser efficacy depending on whether they are used offensively or defensively. According to US Army Weighted Unit Values, tanks in offensive roles have a weight of 64, but only 55 in defensive roles.[2] Other relevant figures are: armoured personnel carriers (APCs) 13 and 6, artillery 72 and 85, anti-tank weapons 27 and 46, small arms 1 and 1.2. Thus, it seems warranted to refer to offense-prone (-oriented) or defense-prone (-oriented) weapon systems and force structures, the latter encompassing not just weaponry but also doctrine, training, deployment, logistics, etc.

Table 15.1. An Operational Definition of Conventional Stability[3]

The Five Aspects and Criteria of Conventional Stability
- "Preemption Stability": The possible bonuses for preemption must be minimized. This can be achieved by three means, (a) the reduction of the number and relevance of targets for preemption, (b) protection for the remaining, inevitable targets, (c) the limitation of weapon systems that can be used for preemptive strikes.
- "Battlefield Stability", part 1: The exploitation of the advantages of the agressor on the battlefield must be inhibited. This means that especially surprise effects and local superiority must be denied. It can be achieved by limiting offense-prone mobility, massing and momentum.
- "Battlefield Stability", part 2: The structural advantages of the defender must be optimally used. This means that there should be no arms control limitations to non-mechanized infantry, terrain preparation, barriers and mine technology. The same holds true for anti-tanks and anti air-PGMs as long as they are not mounted on platforms.
- "Force-Generation Stability": The potentials for force generation must be controlled and shaped in a way that no significant superiority can be achieved at any time or by any mobilization pattern.
- "Arms Race Stability": The overall military setting must make sure that the acquisition of additional offensive capabilities is by far more expensive than the means to compensate them.

Third, technological progress in the conventional realm tends to favour defense-prone armaments more so than offense-prone armaments. Technological progress since World War II has been fastest in the categories of fire (estimated progress rate 10-100) and intelligence (rate: 10-1000) that tend to be more relevant to the defender, while the categories of protection (rate: 3-10) and mobility (rate: 3-7) most crucial to the attacker showed considerably less progress. This progress rate differential becomes especially salient since it translates into cost differentials.

Albrecht von Müller characterizes CS in terms of five criteria (Table 15.1). A CS regime must:

1. minimize possible preemption bonuses;
2. prohibit the attacker from utilizing surprise attack and local superiority;
3. allow the defender to exploit its advantages such as terrain preparation and cost-effective, defence-oriented modern technologies;
4. provide for force-generation stability;
5. provide for arms race stability.

Table 15.2. The Modernization Concept of "Integrated Forward Defence"[4]

(1) Build-up of a net of light infantry and barrier units approx. 25 km deep with following provisions:
- force strength in peace approx. 40,000; in a defence situation approx. 120,000 men
- regional draft system and organization in home defence regiments
- reservists' training twice a year on weekends
- mobilization within 12-24 hours possible, due to close proximity to defence area and high training level.

(2) A systematic improvement of "Close Interdiction" capabilities
- laying of approx. 2,000 km^2 of sensor net by members of engineers' corps (costs: approx. $100 million for net, approx. $100 million for supplementary equipment)
- procurement of approx. 3,000 simple combat drones (cost: approx. $75 million)
- procurement of further intelligent passive ammunition (guided mines, jumping mines, communicating mine fields, etc. for approx. $125 million)
- procurement of approx. 30,000 mobile mines with launchers (costs: approx. $90 million for mines, approx. $75 million for launchers)

= Total costs approx. $565 million, which means price increases even by a factor of 2-3 still tolerable

(3) Maintenance of mobile, heavy-armoured brigades
- with slightly reduced personnel numbers
- not worn down prematurely in the so-called delay battle
- with a high likelihood of being at the right place at the right time
- which can be optimally used according to their characteristics as highly mobile blocking- and counter-attack forces

These criteria tend to reinforce each other, but some trade-offs undoubtedly exist.

15.2 Specific Features of a Conventional Stability Regime

More specifically, von Müller proposes to implement a stability regime as a combination of force restructuring (Table 15.2) and arms control (Table 15.3) measures. Force restructuring would put heavy reliance on light infantry and barrier units in forward positions, and would be complemented by improved and cost-effective "close interdiction" capabilities. Mobile armoured forces would be maintained as operative reserves against break-throughs and for counter-attack but at reduced force levels. Conventional arms control would focus on fairly low ceilings for major weapons platforms (main battle tanks, infantry fighting vehicles, artillery, strike aircraft and

Table 15.3. Outlines of a Stability-Oriented Arms Control Regime for Europe[5]

I. Substantial Provisions

(1) a low ceiling for MBTs (e.g. 10,000) combined with a density limit (e.g. 500 MBT per 100 * 100 km^2);
(2) a comparably low ceiling for heavy artillery (calibres above 100 mm) and rocket launchers, again combined with a density limit that denies offense capable concentrations;
(3) a low ceiling for strike aircraft (e.g. 500) and for armoured helicopters (e.g. 500);
(4) a range limitation for conventional rocket and all other unmanned weapon systems (e.g. 50 km);
(5) geographical limits on ammunition stockpiles (e.g. not closer than 150 km to the border) and forward deployed mobile bridging equipment at all;
(6) logistical infrastructures that require frequent back-up by non-mobile service stations and other installations

II. The Role of Nuclear Forces

For the time being it is probably unavoidable to retain nuclear weapons as a factor of ultimate incalculability. But given a regime of conventional stability, the nuclear component could be substantially reduced.
A limit of 500 warheads for land and air-based systems in Europe might be sensible, of which only 100 would be allowed to be mounted on rockets (with less than 500 km range according to the INF Treaty). The arms control proposals for Europe should be complemented by such a nuclear component.

(All limitations apply to Atlantic to Urals; regarding (1) and (2) there is a sub-ceiling of 50% for the Central Region)

---

armoured helicopters) with a subceiling for the Central Region. In addition, range limitations for rocket systems, density limits for the remaining ground-based heavy platforms, and other constraints on infrastructure and logistics, are proposed. Land and air-based nuclear systems would be limited to 500 warheads. There would be no limitation on defence-oriented "close interdiction" capabilities.

A scheme along these lines would have a range of ramifications: It would trade mobile armoured forces for less mobile but nonetheless highly flexible and well-equipped infantry. Roughly, the aim might be a 1 to 2 ratio of armoured versus infantry and barrier units. If realized on both sides, concerns about maintaining an operative minimum and reservations about far-reaching arms control would become spurious. There would be less of a monoculture than the present heavy reliance on tanks. It would strengthen "denial" capabilities and, thus, de-emphasize the role of nuclear deterrence in Central Europe. Sufficiently strengthening "close interdiction" would alleviate the need for deep interdiction. Some missions air forces, especially for offensive counter air, close air support and air interdiction, would be significantly reduced. Accordingly, time critical and valuable targets would be removed and, hence, preemption becomes less lucrative.

It is less obvious how the criteria of force generation stability would be affected. If, as contemplated, the bulk of the (forward-based) light infantry forces is made up of reservists, early mobilization in times of crisis would be required with probably destabilizing consequences. On the other hand, since the attacker would be much less capable of massive and surprise attack due to lack of (mobile) forces, crisis stability would be increased.[6] By and large, since the threat potential would decline, militarily relevant crises in Central Europe seem likely to diminish.

Ramifications also exist for domestic, alliance and international politics. From the West German perspective, a CS regime as outlined would mitigate serious public acceptance problems, such as the nuisance posed by low-flying aircraft, or the extension of the draft, since air force missions and manpower requirements would change. If the cost estimates as a ballpark figure - allowing for a considerable margin of error - are correct, substantial savings would be accrued compared to the cost of present defence planning. This would ease the strain on national budget problems, perhaps on intra-alliance burden-sharing issues, and probably foster improved international relations. If "defensive" capabilities are cheaper than "offensive" ones, the arms race will tend to slow down and, thus, a major source of East-West confrontation will dry out. Of course, the proposed regime squares well with Eastern proposals made in the past 2-3 years, including Gorbachev's recent announcement for unilateral withdrawals and reductions.

There are some question marks regarding the effects on NATO. On the financial side, the effects could go both ways. Declining military expenditures would ease the common burden, conceivably mitigating the burden-sharing issue. But it is equally plausible that burden-sharing would become even more salient if savings would accrue unequally among the NATO partners. Much will depend on how the new conventional force posture will affect the respective allied contributions in the FRG. The Bundeswehr is likely to provide the lion's share of the infantry troops up front, while the armoured forces further back will continue along the lines of the current corps sectors - albeit in lower numbers and adapted to the new situation. In order to serve NATO support, it would be helpful not to alter the allied deployment pattern too drastically, and to devise intra-alliance mechanisms for mastering both the military and financial transition. Similar deliberations apply to the WTO countries.

With regard to the negotiating process, the CS regime as outlined obviously requires much more elaboration. The most crucial areas seem to me the transition from today's posture, non-reduction constraints, modernization and verification. Given the MBFR experience, procedures must be found to avoid a prolonged data dispute - unlike MBFR, not only in quantitative but also in qualitative terms. The WTO's declared openness to extensive data exchanges and on-site inspections certainly portrays the much heralded new thinking. It seems paramount to focus on common (equipment) ceilings rather than reductions or manpower as the unit of measurement. In order to be attractive to the numerically superior side, the ceilings should be so low as to require more than merely token reductions from the numerically weaker side. In addition, since we are unlikely to see reductions to the proposed low ceiling occur in one step, the transition stages must provide incentives to both sides.

As part of his CS scheme, von Müller proposes the "retrograde ceiling definition" approach as a transition formula. It goes as follows:

1. both sides agree on common ceilings for the weapon categories to be reduced;
2. both sides agree on the number of steps for the reduction process;
3. both sides declare their present holdings;
4. at each step, both sides cut their excess holdings (above the common ceilings) to one half.

The intriguing idea of this approach is that it circumvents the initial data problem. Provided that strict verification of the remaining holdings is possible at each intermediate ceiling, the approach has a built-in disincentive to cheat on one's initial holdings: the lower the initial figure, the lower the intermediate ceilings. Yet, without strict verification, it would "pay" to underrate one's holdings. In addition to ceilings, intermediate limits to the permissible density of equipment deployments may be needed. For practical purposes, a few (fairly large) intermediate steps are preferable to a more gradual formula. In any case, the algorithm assures that in each step the asymmetries are further reduced until parity is reached in the final step.

15.3 Other Measures

A plethora of "non-reduction constraints" has been proposed by various authors. Such measures could be part of a CFE agreement (especially in the context of a verification regime), part of the second phase CDE talks, or be promulgated unilaterally. The measures include: consultations on doctrine and strategy; areas free of, or with limited numbers of treaty-limited items; limitations to exercises, manoeuvres, and out-of-garrison movements of armoured and mechanized forces; restrictions on key items for offensive operations, such as bridging or mine-clearing equipment, withdrawal of ammunition stockpiles to backward locations, "defensive" logistics, etc.; open-sky policies for selected areas allowing monitoring from air on a permanent basis; etc.

A particularly salient area - albeit difficult to implement - would be restrictions on modernization. Eastern and Western analysts have proposed caps on modernization and compensations - for example, prohibiting the introduction of new platforms, or equipping existing ones with new types of weaponry. While most experts consider this a good idea in principle, they point to the substantial definitional and verification problems. Some maintain that a cap on modernization is superfluous if there are stringent limits to the number of platforms. Others argue that restrictions should only apply to such items that endanger conventional stability. This might be particularly pertinent for aircraft. Defensive systems, on the other hand, should be technologically state-of-the-art.

Rapid advances in basic technologies (electronics, optics, sensors, micromechanics, materials, etc.) make a distinction between military and non-military technology increasingly spurious. Many basic components are clearly dual-use. Capping modernization with respect to these components will be unrealistic. This appears to be different in the case of major platforms and their equipment. If new physical principles are to be applied, challenge inspections should be able to detect them. Definitional problems should not be insurmountable, given the experience with the ABM and INF Treaties - if the political will for modernization caps is there! If "standard" physical principles applied to new techniques and equipment are the issue, then tight controls on testing should be able to give sufficient assurance that only permitted technologies/parameters are used. No major improvement can be introduced without extensive testing. If a modernization cap would be agreed upon, then test sites would be crucial targets for inspections.

Basically, a CS regime would be verifiable if the contracting parties are willing to live up to their pre-negotiating rhetoric.[7] Both sides seem to be in agreement on the essentials of a verification system, e.g., mandatory on-site inspections on demand in addition to routine inspections, non-interference with national technical means, additional confidence and security-building measures, permanently-manned check points, designated storage sites for equipment. Some elements of it might be put in place even prior to the beginning of the formal negotiations. Generally speaking, the objective of a verification regime should be to ascertain compliance on a militarily relevant scale rather than the detection of any single violation.

Besides verifying compliance with common ceilings, and possibly transitory or spatial sub-ceilings, verification of density limits would be a major task. Permanent observers at the division or even brigade or battalion level would greatly facilitate verification. In any case, designated secured and mutually controlled storage sites for treaty-limited equipment will probably be indispensable. New schemes involving tags and seals are being developed that increase the risk of detection of deliberate non-compliance.[8]

15.4 Notes and References

1 The author made his presentation in lieu of Albrecht von Müller from the same institute and has incorporated von Müller's ideas at central places throughout this paper

2 William P. Mako, U.S. Ground Forces and the Defense of Central Europe, Washington D.C.: The Brookings Institution, 1983

3 Albrecht von Müller, Towards Conventional Stability in Europe, Proc. 38th Pugwash Conference on Science and World Affairs, Dagomys, USSR, 1988

4 von Müller (note 3)

5 von Müller (note 3)

6 For similar trade-offs, see: Karsten D. Voigt, Konventionelle Stabilisierung und strukturelle Nichtangriffsfähigkeit: Ein systematischer Vergleich verschiedener Konzepte, Aus Politik und Zeitgeschichte, no. B18, April 29, 1988

7 Jonathan Dean, The New NATO-Pact Force Reduction Talks - An Optimal Outcome, Paper for the American Academy of Arts and Sciences, Project on Conventional Forces in Europe, May 18, 1988

8 Klaus Jacob, Aspects of the Verification of Conventional Arms Control Measures in Europe, this volume, Ch. 19; Requirements of a tag for conventional arms control, AG I (88) D/10, Paris: Western European Union, 15 July 1988; Richard Garwin, Tags and Seals for Verification, The Council for Arms Control Bulletin, no. 40, October 1988

Part VI

Confidence and Security-Building Measures in Europe

16. Verification of Confidence and Security Building Measures: Evolution and Future Prospects

Bernd A. Goetze[1]

16.1 Evolution of CSBMs

Although various forms of Confidence-Building Measures (CBMs) have been a part of the international political landscape as far back as the 1648 Treaty of Westphalia,[2] it is their more recent evolution in the context of the Conference on Security and Cooperation in Europe (CSCE) which continues to be of interest to those concerned with the management of European, and by extension, Superpower security relations.

From their modest beginnings in "Basket I" of the 1973-1975 CSCE negotiations in Geneva and Helsinki, these measures have evolved from rather vague, voluntary undertakings in the Helsinki Final Act, to an interacting set of substantive obligations in the Concluding Document of the Stockholm Conference on Confidence and Security-Building Measures (CSBMs) and Disarmament in Europe (CDE), which came into force on the 1st of January 1987.

The CSCE-CSBMs road from Helsinki to Geneva, from Belgrade (1977-78) to Madrid (1980-83), and from Stockholm (1984-86) to Vienna (1986-88), has been difficult and often barren, as it twisted through the rise and fall of tensions in East-West relations. But it is useful to recall that it was in these negotiations that the first significant breakthrough in verification modalities came; it was in these negotiations that the concept of "on-site" inspection was finally given practical expression; and it was as a result of these achievements that the first tentative steps towards dispelling the fear, mutual suspicion and lack of confidence in the intent of major powers arising from the massive concentration of armed forces in Europe, were taken.

16.2 The Helsinki Regime

The original 1973-75 CSCE negotiations (among the 35 countries consisting of all the states of Europe (less Albania) plus USA and Canada) related to military aspects of security were conducted against a background which saw major East-West arms control issues discussed in more narrowly constituted fora such as MBFR and SALT. Nevertheless, the CSCE Final Act's contri-

Table 16.1. Helsinki CSBMs regime[3]

Threshold of notification:	25,000 troops
Type of forces:	Independent ground forces; independent amphibious forces; independent airborne forces; combined forces in which land forces exceed 25,000 troops
Advance warning:	21 days
Area of notification:	The whole of Europe and, if applicable, the adjoining sea area and air space (general regime). Special regime for countries whose territory extends beyond Europe (USSR and Turkey), an area within 250 km from the frontier facing or shared with a European participating State.
Form & contents of notification	No standardization; limited contents.

Table 16.2. Final Act CSBMs - The Record 1975-86 [4]

	West	East	NNA	Totals
Notifications	77	32	21	130
Invitations to Observers	50	10	12	72

bution in this area was significant in that it formalized the concept of reducing the risk of misunderstanding of the purpose of routine military activities through the introduction of certain limited CSBMs.

These measures included the prior (21 days) notification of major military manoeuvres "exceeding a total of 25,000 troops ... which take place on the territory, in Europe, of any participating States as well as, if applicable, in the adjoining sea area and air space;" discretionary notification of smaller military manoeuvres; and exchange of observers to attend military manoeuvres (Table 16.1).

The participating States also expressed an intention to review at a later time "the question of prior notification of major military movements" and agreed on the desirability of promoting "exchanges by invitation among their military personnel, including visits by military delegations".[5]

Agreement on these voluntary measures was a constructive first step, going beyond a mere restatement of general arms control and disarmament objectives. While their importance to confidence-building could ultimately only be judged by their record of implementation, the care with which these early CSBMs were negotiated is testimony both to the substantial interest which a majority of CSCE states had in such measures and to the real difficulty of proceeding to far-reaching security arrangements in the CSCE context.[6]

Being part of the Final Act, these CSBMs entailed a political commitment undertaken at the highest level but imposed no mandatory obligations. The overall spectrum of this political undertaking extended from the declaration of participating states that they "will" notify other states of their major military manoeuvres exceeding 25,000 troops, but with the important qualification that the measure "deriving from political decisions rests upon a voluntary basis ...,"[7] to the mere acknowledgement that they "may at their own discretion" notify their major military movements.

Bearing in mind this important variation, an examination of the 1975-86 implementation record of Western, Eastern and Neutral and Non-Aligned (NNA) participating states (Table 16.2) provides no conclusive evidence that any state or group of states failed to provide notification of a major military manoeuvre. Although the then fifteen NATO countries conducted a greater number of notifiable manoeuvres (77) when compared to the seven Warsaw Pact countries (32), the West provided a much more expansive range of on-site verification opportunities by inviting observers to more than twice as many of its exercises (65% vs. 31%).

In most Western cases these invitations were addressed to all CSCE participants, or to a large proportion of them, including Warsaw Pact states. Generally, observers to NATO manoeuvres were given ample opportunity to carry out their tasks by providing them with detailed briefings, experienced escorts, fixed and mobile observation posts, and by allowing visits to the actual exercise area, contacts with command elements and troops in the field. Use of binoculars, and in most cases cameras, was also permitted.

NNA countries implemented the security provisions of the Final Act with evident good will. A total of 21 manoeuvres were notified and a substantial number of observers were invited representing Eastern, Western and other NNA interests. These observers were accorded privileges similar to those extended by Western CSCE states and were thus able to conduct their observations in a satisfactory manner.

As regards observers invited by Warsaw Pact states, they were generally not given adequate facilities to enable them to carry out their tasks effectively. In many cases, total observation time was only a few hours, briefings were vague, observers were not allowed to move freely, and were shown set-piece demonstrations rather than actual simulated battlefield manoeuvre activity.

These restrictions not only prevented observers from gaining a clear understanding of the purpose of the manoeuvres, but made it virtually impossible to formulate any realistic appreciation of their size and scope, or of the capabilities of the troops involved.

The absence of binding obligation in the Final Act CSBMs, together with the restrictive Warsaw Pact application of the observer provisions, severely limited the confidence-building value of the Helsinki measures. However, in implementing these CSBMs, Warsaw Pact states have displayed greater openness about their military activities, thereby meeting at least to some extent one of the original Western objectives in the CSCE.

The political value of the Final Act CSBMs, as modest instruments advancing the cause of détente in Europe, cannot be denied. However, their voluntary character and the absence of any substantive verification provisions made them unreliable as indicators of a potential aggressor's intentions in times of crisis or tension. While this reality depreciated their security and confidence-building potential, the Helsinki CSBMs played a critical role in setting the stage for the subsequent negotiations of a politically and militarily more meaningful regime in Madrid and Stockholm.

16.3 The Stockholm Regime

At the time of the signature of the Final Act it was considered that experience gained in the implementation of its provisions could lead to developing and enlarging measures aimed at strengthening confidence. The Belgrade Follow-up Meeting (1977-78) failed in such a further development of CSBMs, although a review of the record gave participants more than sufficient reason to do so.

Based on that opening in the Final Act, and with a view to the 1980 CSCE Follow-up Meeting (FUM) in Madrid, France took the initiative in 1979 to develop a mandate for a post-Madrid Conference on Disarmament in Europe (CDE) which in its first phase would adopt a system of improved CSBMs.

To achieve their purpose, the French argued that the improved CSBMs must be as objective as possible, and have a definitive beneficial effect in terms of the military security situation in Europe. To this end, they would have to permit public knowledge of military activities; take account of the geographical realities of Europe; cover military activities considered as being a potential threat to participating states; not be confined to notification procedures; be applied on an obligatory basis, and - most importantly - be subject to adequate arrangements for verification.

Incorporating the key elements of this French proposal, the Western allies went on to define specific criteria for future Confidence-and-Security-Building Measures (CSBMs) which served as the basis for the Stockholm conference mandate negotiated in the course of the 1980-83 Madrid FUM. In addition to insisting that new measures be obligatory, be militarily significant, and apply to the whole of Europe from the Atlantic to the Urals, the West demanded that they be verifiable to the extent that non-compliance with an agreed measure would be detectable early enough to allow for rapid resolution of a possible misunderstanding, or for an appropriate defensive response should that become necessary.

Recognizing that the degree to which compliance of obligations can be verified determines the level of confidence which can be generated, the verification provisions agreed to in the Stockholm regime are obviously a quantum leap forward from those in the Helsinki Final Act (Table 16.3).

The combination of the obligation to accept observers at notified military activities, with the right to conduct on-site inspection in order to dispose expeditiously of concerns about observed or monitored military activities, which, in the opinion of the requesting state pose a threat or may have violated an agreed threshhold or constraint, is the crux on which the success of the Stockholm CSBMs rests.

The very essence of confidence-and-security-building is at stake here. The lack of adequate verification in past arms control and disarmament agreements has so emasculated their credibility that states have simply not been willing to put their fundamental security interests at risk in an arrangement which did not guarantee their ability to verify compliance with the terms of an agreement. Therefore, the implementation and verification of CSBMs agreed by the parties to the Stockholm Concluding Document will be a marker not only for the continued viability of the CSCE confidence-building process, but could also serve "as an underpinning for more conventional type of arms control agreements and as an alternative when traditional types of agreements prove impossible to negotiate or sustain."[8]

As noted earlier, the Stockholm regime came into effect on 1 January 1987. Since that time the record shows that, with some minor deviations, all participating states have essentially complied with the provisions of the new CSBMs. With two exceptions (Austria and Yugoslavia were late in notifying activities which took place on their territory), basic thresholds for notification of military activities have been adhered to, even if in some cases NNA and Warsaw Pact countries omitted certain technical details, such as start and end point of manoeuvres, exercise designators, breakdown of unit and major equipment categories, and level of participation by forces of other nations in joint exercises.

For example, all forecasted activities which were included in the annual calendar for 1987 were in fact notified, with the exception of one French, one British and one Swiss manoeuvre, which in the event fell below the verification threshold specified in the Stockholm agreement. The United States (and the FRG as host nation), notified two land force exercises which had been included in their annual calendars but in practice involved less than 13,000 troops. In the same vein, Hungary notified Exercise BASALT 87 which was composed of only 8000 ground forces, and the Soviet Union provided voluntary notification of their participation in two joint activities on the territory of Hungary and the CSSR respectively.

Turning to the central issue of CSBM observation and verification, the practice to date is clearly encouraging and gives rise to optimism for the

Table 16.3. Stockholm CSBMs regime[9]

	Notification
Type of forces:	"Military activities in the field conducted as a single activity", national or multinational - i.e. ground forces (independent and combined), amphibious and airborne forces, transfers and concentrations, alert exercises (Abolition of the distinction between manoeuvres and movements)
Threshold of notification:	- 13,000 ground troops or 300 battle tanks if organized into a divisional structure or at least two brigades/regiments. Sub-threshold for air forces: 200 sorties by aircraft, excluding helicopters. - 3,000 amphibious or airborne troops
Advance warning:	42 days - except for alert exercises, notifiable at once
Area of notification:	The whole of Europe as well as the adjoining sea area and air space (no more special derogation for the USSR and Turkey)
Form & contents	Standardized form; extended contents of notification
	Observation
Threshold of observation:	17,000 troops (ground) or 5,000 troops (amphibious airborne)
Program of observation:	Standardized
Exceptions:	- No observations of restricted locations, installations or defense sites - No observations of alert activities having a duration less than 72 hours

Annual calendars

Exchange (each 15 November at the latest) of forecasts of notifiable military activities for the subsequent year, under a standardized form

Constraints

Time and size constraints: two years of advance notice for any notifiable military activity beyond 40,000 troops, prohibition of military activities beyond 40,000 troops and under 75,000 unless notified one year in advance; prohibition of military activities beyond 75,000 troops unless notified two years in advance, unforecasted military activities should be "as few as possible"

Verification

- Use of National Technical Means (satellite reconnaissance)
- On-site inspections; compulsory (no room for refusal); short notice (24 hours) and duration (48 hours); Inspections on the ground, from the air both, by a maximum of four inspectors; passive quota (three inspectors per year from different States other than members of the same military alliance); access, entry and unobstructed survey except for restricted areas whose number and extend should be "as limited as possible"

future. There have been no indications of substantive violations of the requirements to invite observers to designated activities, nor have any on-site inspection requests been refused. NNA countries did not conduct observable manoeuvres in 1987. Reports of observers seem to indicate that, although East-West differences similar to the pattern experienced under the Helsinki regime remain, in general terms host nations have approached the work of observers in a professional, non-confrontational manner. Differences in the structure of Warsaw Pact and Western manoeuvres still create the impression that many observed Eastern activities were staged more for the benefit of the observers than for the training of participating troops.

In keeping with a narrow interpretation of the Stockholm provisions, Warsaw Pact countries permitted observers to carry binoculars, but tape recorders and cameras were not allowed in exercise areas. Briefings tended to be rather superficial and maps provided were often of such large scale (1:500,000) that details of the exercise were difficult to relate to the ground. As a result, observers were for the most part unable to assess accurately the full scope of the activity in relation to its notification content, nor to follow the sequence of events as the various phases of the exercise unfolded.

Early indications of 1988 experience are that efforts are being made to improve working conditions of observers, and that greater flexibility is being demonstrated in meeting requests for observation of specific activities (e.g. river crossings) from both ground and air platforms. The further expansion of facilities and permissible activities by observers at notifiable military events should be a major element of any future negotiation to enhance the existing regime. That having been said, it is extremely important not to minimize the progress which has been made since Helsinki, and to recognize the positive atmosphere prevailing in this area.

The changing East-West political climate and the widening effects of Glasnost and Perestroika in Eastern Europe have undoubtedly been catalytic factors in this development, and that in itself augurs well for the future.

Turning to the record of on-site verification when "compliance with the agreed CSBM is in doubt,"[10] an equally positive preliminary assessment appears warranted. Since the Stockholm verification arrangements came into effect on the 1st of January 1987, no CSCE state has refused a request for an on-site inspection and, in fact, a total of 18 such inspections have been carried out: nine by Western countries and nine by members of the Warsaw Pact (Tables 16.4 and 16.5). The Soviet Union has conducted five inspections: two in the FRG; and one each in Turkey, Norway and the United Kingdom. The United States has executed five too; two in the USSR, and one each in Hungary, Poland and the GDR. The United Kingdom has carried out two: one in the Soviet Union and one in the GDR. The GDR has also carried out two, both in the FRG. For the rest, Turkey has conducted one in

Table 16.4. Stockholm CSBMs - The Record 1987-88

	West	East	NNA	Totals
Notifications	30	47	8	85
Invitations to Observers	17	15	3	35
On-Site Inspections	9	9	0	18

Table 16.5. CSBMs Verification - Inspection Matrix (1987-1988)

USSR	USSR	USSR	USSR	USSR	
	TUR	FRG	NOR	FRG (US)	UK
USA	USA	USA	USA	USA	
	USSR	HUN	GDR	POL	USSR
UK	UK				
	GDR	USSR			
GDR	GDR				
	FRG	FRG			
FRG					
	GDR				
TUR					
	USSR				
BUL					
	ITAL				
POL					
	FRG				

Host		Insp
USSR	4	5
USA	0[a]	5
UK	1	2
GDR	3	2
FRG	5[a]	1
Turkey	1	1
Bulgaria	0	1
Norway	1	0
Poland	1	1
Hungary	1	0
Italy	1	0

[a] US Forces have been inspected in course of verification in FRG

the USSR, Bulgaria one in Italy, Poland one in the FRG, and the FRG one in the GDR. None have thus far been initiated by NNA states.

In all cases notice of intent to conduct an inspection was given in strict accordance with the parameters set out in the Stockholm Document. Facilities provided to inspectors by the challenged state were adequate for the mission, and a businesslike atmosphere generally characterized the proceedings. Warsaw Pact inspectors appeared particularly well equipped with maps, detailed information about the activity to be inspected, and a very clear idea of what they wanted to see. Western teams were treated correctly and encountered no insurmountable obstacles in conducting their inspections. Notwithstanding some unsubstantiated concerns raised by the Soviet Union, in connection with the level of allied participation in exercises they inspected in the FRG, no questionable practices or outright breaches of agreed CSBM provisions were discovered by any party as a result of these inspections. With these reassuring results as a backdrop, one could almost form the impression that this initial series of inspections had more to do with simply exercising the right to do so and working out procedural anomalies than with actual suspicions of CSBM violations. If that is the case, then the effectiveness of the Stockholm measures may have exceeded our initial expectations.

One worry that has appeared in some circles is the risk of inspections being abused for intelligence gathering purposes, or to give effect to other political objectives (e.g. recognition of Berlin as the capital of the GDR). Such possibilities are of course inherent in any no-refusal challenge verification arrangement. Preliminary analysis of inspections carried out by Warsaw Pact, particularly Soviet teams, lend a certain credence to such worries. However, when taken on balance, and allowing for the fact that all parties are in the process of adjusting to an unprecedented level of intrusiveness in this verification measure, the constructive and non-polemical manner in which it has been implemented up to this point is nothing short of remarkable. In short, the benefits at this stage seem to outweigh the liabilities.

16.4 Future Prospects

As the CSCE Follow-up Meeting in Vienna gradually draws to a close, the expectation is that as part of a balanced concluding document, participating states will mandate a new round of negotiations dedicated to improving the confidence and security-building process begun in Helsinki and advanced so markedly in Stockholm.

Without compromising the criteria for new measures agreed in Madrid, it should be possible to build on the Stockholm record. In the area of information measures, one could envisage enhancing transparency through agreement on regular exchanges of basic data about the location, strength, equipment, and organization of major military units (Division above) within the CSBMs area of application. Discussion might also focus on exchange of information on military expenditures and, as the CSBM regime matures, developing a suitable communications network for the speedy passage of notification messages, assistance in quickly clearing up misunderstandings and - most importantly - be available for crisis management if violations are suspected.

Notification measures could benefit from adoption of the "out-of-garrison" concept, so that any military activity regardless of its purpose would be subject to notification when its size or scope exceeds agreed thresholds or other parameters. In this context, the lowering of the threshold for notification and observation to the level of 10,000 troops or one Division equivalent would contribute to lowering the fear of surprise attack, and could later be integrated with whatever accompanying measures may be negotiated in a future East-West conventional stability agreement.

The declared intentions of NATO and Warsaw Pact countries in launching a new negotiation, designed to remove the current asymmetries in their conventional forces in Europe, mitigates against any attempt at this point to incorporate additional constraint measures into the existing CSBM regime, as this would risk working at cross purposes in two distinct fora. However, should the Negotiations on Conventional Armed Forces in Europe (CFE) fail to produce the desired results in a reasonable timeframe, then a fresh look at CSBMs, designed to constrain certain types of major military activities and their location, could be considered as an interim measure.

With respect to verification of CSBM compliance, improvements are necessary to ensure that observers are able to evaluate the extent to which observed activities conform with the content of the notification. They must be able to understand the aim and scope of the activity, have the necessary freedom of movement to be able to observe its evolution, have sufficiently detailed maps, and be able to communicate freely with various echelons of participating forces.

The existing on-site inspection procedure appears to be working reasonably well thus far. As more experience is gathered, procedural adjustments will suggest themselves which should not prove too difficult to incorporate, particularly if the current favourable trend in East-West relations is sustained over time.

The willingness expressed in the 29 October 1988 Warsaw Pact Foreign Ministers Budapest statement on CSBMs in Europe, to work towards "improving conditions for inspections and working opportunity for observers", could contribute to a constructive climate for the resumption of negotiations.

The conclusion of the bilateral INF arms control treaty between the United States and the Soviet Union, the prospects for an even more far reaching START agreement, and the initiation of Negotiations on Conventional Armed Forces in Europe between the members of NATO and the Warsaw Pact, place ever increasing demands on the efficacy of negotiated verification arrangements. The CSCE confidence-building process has been an important contributor to the evolution of such arrangements. Therefore, as participating CSBM states reconvene after the conclusion of the Vienna Follow-up Meeting, in order to build on the achievements of Helsinki and Stockholm, there is no reason to believe that further advances will not be possible. The future security of Europe depends on it, and the peoples of the CSCE states will accept no less.

16.5 Notes and References

1 The views expressed in this paper are those of the author and do not necessarily reflect those of the Government of Canada or any of its sub-ordinate Departments or agencies

2 For an excellent summary of pre-Helsinki CSBMs see: James Macintosh, Confidence (and Security) Building Measures in the Arms Control Process: A Canadian Perspective, Ottawa: Department of External Affairs, August 1985, pp. 16-26

3 International Defence Review, no. 10, p. 7, 1988

4 International Defence Review, no. 10, p. 8, 1988

5 The Final Act, Helsinki: Government of Finland, August 1975, pp. 10-12

6 L. V. Ferraris (ed.), Report on a Negotiation, Geneva: Sijthoff and Nordhoff, 1979, pp. 179-204

7 See: Final Act (note 5), Document on CSBMs and Certain Aspects of Security and Disarmament

8 J. Alford (ed.), The Future of Arms Control, Part III, Confidence Building Measures, Adelphi Papers, no. 149, p. 1, 1979

9 International Defence Review, no.10, p. 9, 1988

10 Paragraph 66 of the Stockholm Concluding Document

17. Observations of Military Exercises and On-Site Inspections

Manfred Graczynski

17.1 Introduction

The successful implementation of the INF Treaty and the intensive negotiations on disarmament in other fields, the multiple efforts aimed at the settlement of regional conflicts made it possible that there has been a turn for the better in international relations.

The understanding is gaining ground that more weapons do not allow for more security and that military superiority over the other side is not achievable. At the same time, both sides are gathering new experience in eliminating weapons and in implementing confidence and security-building measures under strict international control. But we cannot yet take it for granted that peace is secure. There is no other region in our world where such vast numbers of means of mass destruction are being stored as in Europe, and where armed forces, counting millions of personnel, of the most important military alliances - NATO and Warsaw Treaty - face each other.

For this reason it is necessary to make use of all existing possibilities to initiate a turn towards stable détente, to overcome gradually military confrontation, with the aim of enhancing comprehensive cooperation between East and West. Confidence and security-building measures are a stimulating factor to solve these problems.

The implementation of the Treaty between the Soviet Union and the United States on the elimination of their intermediate-range and shorter-range missiles is being watched by public opinion with extreme interest and greeted almost unanimously.

The German Democratic Republic undertook all efforts in order to support actively the preparation and implementation of this Treaty. Even before the INF Treaty came into power, the withdrawal of the Soviet SS-12 missiles (OTR-22) deployed in the GDR was terminated on March 21, 1988 as result of an agreement between the GDR and the Soviet Union. All over the world, this step was appreciated as an expression of the striving for disarmament, as an important confidence-building measure and good-will-gesture contributing to a more healthy international situation.

In this context it may be of interest to note that in the former Soviet missile base Warenshof which was handed over to the GDR trade unions, more than 800 people may spend their vacation since July 1, 1988. This testifies to the benefit which disarmament can bring along.

Late in October 1988, an inspection team of the United States has inspected former deployment sites in the GDR, following the provisions of the INF Treaty. They monitored the complete withdrawal of Soviet SS-23 missiles (OTR-23) from Soviet military installations in Jena and Weissenfels. With this, the GDR has once again fulfilled its obligation to support the United States in conducting inspections on its territory.

It is evident that the thorough and strict fulfillment of all obligations resulting from the agreed inspections by all sides enhances confidence on our continent.

17.2 The Stockholm Document

The Stockholm Document which has been in power since January 1, 1987, constitutes a basis for the obligatory fulfillment of confidence and security-building measures, departing from a contractually fixed agreement.

The provisions agreed upon include:

- the obligation of prior notification, as well as invitation of all participating states to observe certain military activities; and
- the right to address a request for inspection to another participating state, when the compliance with the agreed confidence and security-building measures is in doubt.

The importance of these measures for a more healthy political climate reveals itself as the provisions of the Stockholm Document are being implemented.

Up to now, approximately 240 generals, officers and diplomats from more than 20 CSCE-States participated at the invitation of the GDR Government as observers in 6 military exercises on its territory. Based upon the prepared programme the observers could convince themselves that the notified military exercises, corresponding to the military doctrine of the Warsaw Treaty member states, were of a defensive character and thus did not threaten any state.

During the exercises the observers were given extensive possibilities to visit positions of troops and zones of activities; to communicate with commanders and troops; to get close to the equipment and armament; and to use their own photo cameras and dictaphones.

During their stay in the GDR, the observers made a positive assessment of the possibilities granted for the observation. All observers were unanimous in saying that during all their stay on GDR territory there prevailed an open-minded and frank atmosphere; that there was a climate favouring talks and contacts; and that real life was put into the Stockholm Document.

In order to verify compliance with the provisions of the Stockholm Document, three inspections have been conducted on the territory of the GDR. Inspection teams from NATO countries, each of them composed of four officers, could within 36 hours after the request was issued, and for a period of 48 hours during ground and aerial inspection, gather comprehensive information on compliance with the relevant provisions of the Stockholm Document.

The conclusions derived from the inspection reports, which were handed over to all CSCE-participating states, testified that no provision of the Stockholm Document was violated.

GDR press, radio and television informed broadly on the course and the results of the exercise observations and the inspections.

In general, we can state that the adoption of the Stockholm Document was an important step because such obligatory measures like prior notification, observation and inspection of military activities from the Atlantic to the Urals give a practical example of how the turn from confrontation to cooperation can be set into motion.

17.3 NATO Exercises

The measures agreed upon in Stockholm as such mean neither disarmament nor arms reduction, but they can and must precede or accompany them.

So far, prior notifications and annual calendars of military activities have not led to changes in NATO proceedings for the conduct of large scale military activities. During the NATO exercise series "Autumn Forge", in military manoeuvres covering an area from the polar circle up to the Mediterranean, 500,000 troops were trained this year at very close proximity to the border with the Socialist States.

In our view such kind of military activities is not confidence-building and gives us cause for concern, since from these exercise situations aggressive operations may develop quickly.

17.4 Need for Extending of CSBMs

It is becoming clear that the measures agreed upon in the Stockholm Document are limited in their military importance as well as in their political evaluation. Under the Stockholm Document only certain activities of the land forces are to be monitored.

The risk of a surprise attack is not considerably reduced, because independent activities of the air and naval forces are not covered by the verification procedures. It goes without saying that the air forces in terms of deployment possibilities, speed and manoeuvrability is superior to the land forces. Attack aviation is capable of delivering surprise attacks on strategic targets that can be engaged within one and a half up to two hours to a depth of 1,000 km or more.

The naval forces with their aircraft carriers,battle ships and landing ships represent a large offensive capability. Each aircraft carrier is a sea-going base for about 100 aircraft, with approximately 40 of them nuclear-capable. Out of the 15 US aircraft carriers, 7 are attached to the Atlantic Fleet.

Taking this into consideration, new confidence and security-building measures should be developed with the aim of including, eventually, the activity of all services. Only in this way will the risk of a surprise attack be reduced and covert preparations for an aggression be excluded to a large extent.

The negotiations on confidence and security-building measures in Europe could soon be continued with the aim of:
- reducing the risk of an armed conflict in Europe;
- reducing and eventually eliminating once and for all the risk of a surprise attack;
- strengthening common security;
- giving the military activity a strict defensive character;
- increasing openness and calculability; and
- supporting practical steps to implement disarmament measures.

Starting from the experience obtained with the implementation of the Stockholm Document, it seems feasible to proceed to an agreement of a new generation of confidence and security-building measures. These could include, first of all, measures of a constraining character, for example:
- to limit the scale, quantity, duration and frequency of exercises and to prohibit large-scale exercises (e.g. with participation of more than 40,000 men) and transfer of equipment over large distances;
- to limit alert exercises;
- to refrain from exercises at the immediate vicinity to borders in Central Europe, e.g. within a strip of 50 km along the contact line between Warsaw Treaty and NATO; and
- to include exercise series with prior notification.

At the same time it is becoming ever more evident that such confidence-building measures must be envisaged as:
- prior notification of activities of the air and naval forces, the invitation of observers to assist in these activities and the limitation of such activities following agreed parameters;
- extension of confidence and security-building measures to the territories of all CSCE-participating states;
- establishment of zones of confidence and security in some regions of Europe and in the adjoining sea areas; and
- installation of a centre for the reduction of the threat of war and for the prevention of surprise attacks.

This new generation of confidence and security-building measures could be linked with a broad variety of measures aimed at increasing openness and ascertainment of military activities. These could, among others, include:
- exchange of data on armed forces and their activities;
- creation of permanent joint observation posts at points of strategic importance;

- improvement of conditions for inspections and expansion of working possibilities for observers; and
- exchange of military attachés between all CSCE-States.

Recently, the Warsaw Treaty member states, meeting in Budapest, once again reaffirmed their readiness to continue as soon as possible the negotiations on confidence and security-building measures in Europe with the aim of their enhancement.

Notwithstanding the fact that our objective is solutions covering the whole of Europe, we consider regional solutions highly appropriate to contribute substantially to the building of confidence and to disarmament, and to help in the process of preparing all-European agreements reaching from the Atlantic to the Urals.

The potential of surprise attacks will, without any doubt, be reduced, and mutual confidence, considerably strengthened, if threatening weapon systems will be removed from strips or zones along the line of contact between both alliances. They would create more confidence, and more security, because they are at the sensitive dividing line between the two military blocs.

There is no doubt that mechanisms for peaceful crisis management and the prevention of military incidents in Central Europe, such as "hot wires" between Berlin, Prague and Bonn; centres for the build-up of confidence with the necessary technical equipment; and a joint satellite observation may contribute essentially to confidence-building and give a fresh impulse to the reduction of armed forces and conventional armaments.

It is part of confidence and security-building to compare the military doctrines of both military alliances with the aim of enhancing the strictly defensive outline of both sides' military potentials. In this context the Minister of National Defence of the GDR proposed a dialogue between the Defence Ministers and military experts of both German states on this issue. It is through contacts at this level that, in a competent and realistic way, steps towards the strengthening of security and confidence can be taken at the contact line between both military alliances.

The reduction of military confrontation opens up the real possibility to shape an European system of common security and manifold cooperation.

Recently, the highest representative of the GDR, Mr. Erich Honecker, referred to this issue as follows: "The GDR conscious of her responsibility at the sensitive contact line between the two most powerful military alliances of our time undertakes everything that is within her scope to have a share in shaping a peaceful European house where socialist and capitalist neighbours can live securely, following house rules that are binding for all and where they can build their future in self-determination and without interference from outside." [1]

17.5 Notes and References

1 Neues Deutschland, 15 September, 1988

Part VII

Verification of Conventional Force Reductions

18. Conventional Arms Reduction in Europe: A Verification Model

F. Ronald Cleminson

18.1 Introduction

It now seems certain that two sets of negotiations on arms control matters are likely to emerge from the follow-up meeting of the Conference on Security and Cooperation in Europe (CSCE), now in its concluding days in Vienna. The first is likely to be a continuation of negotiations on Confidence- and Security-Building Measures (CSBMs) following from the Stockholm Conference. It will benefit, in terms of verification, from the experience already gained through the implementation of the observation and inspection provisions of the Stockholm document.

The second set of negotiations will focus on conventional arms reductions in Europe. These negotiations will engage the 23 nations of the North Atlantic Treaty Organization (NATO) and of the Warsaw Pact (WP) in an attempt to achieve enhanced stability in Europe at lower levels of armament. This will necessitate: (a) asymmetric reductions of equipment and manpower in certain areas; (b) reduced levels of equipment with identified offensive capabilities; and (c) a lowering of the capability to initiate a surprise attack. Unlike the CSBM negotiations which seek to build confidence but do not directly affect security, reduction of forces in the second set of negotiations could have a significant, if not an overwhelming, effect on the security of each of the negotiating participants, both individually and as part of the existing alliances. Adequate assurance of compliance will be, therefore, an absolute and essential prerequisite to the achievement of a successful agreement.

The verification challenge which emerges from the Negotiations on Conventional Armed Forces in Europe (CFE) therefore, is to develop a package of verification methods which will be broad enough to meet the compliance requirements of the agreement while stopping short of intelligence gathering. A complicating factor will be the necessity to design a system, multilateral in scope, which can operate effectively. It will be necessary to find the trade-offs required to visualize the concurrent operation of a multilateral verification system which can work with, but not impinge upon, the present National Technical Means (NTM) which are extensively used by the United States and the Soviet Union (for their own national security purposes).

The purpose of this paper is to discuss briefly some of the verification requisites likely to emerge and to suggest a number of verification methods which, in combination, might provide the level of confidence of compliance required by participating states.

18.2 Verification and the European Scenario

In approaching the concept of verifying a conventional armed forces agreement in Europe, we must recognize that much useful data can be extrapolated from the experience of the bilateral agreements, between the United States and the USSR. It is a fact that to date, as a result of a series of significant bilateral arms control agreements, it is the United States and the Soviet Union which possess the bulk of the world's experience with respect to the verification of arms control agreements, particularly using advanced remote sensing techniques including space-based systems. It should be recognized that, with the possible exception of the Non-Proliferation Treaty (NPT) of 1970, no multilateral treaty of sufficient significance yet exists which would warrant the high level of technological sophistication and the financial expenditure required for space-based remote sensing to verify its provisions. One significant result of the INF agreement, will be to make the conventional armed forces balance in Europe of greater significance in maintaining a credible deterrent posture. A sustained negotiation to redress some of the Soviet Union's conventional superiority is now essential. In such a case, the resulting agreement will require the direct input from a significant number of countries in addition to the superpowers, and thus will be multilateral in character. It is likely that nations acceding to such an agreement will insist upon at least the opportunity to participate in the verification process developed to ensure compliance. It is possible that members of the major alliances in Europe might be prepared to accept some sort of alliance to alliance procedure.

Although the two new sets of negotiations in Vienna are to be recognized as separate and distinct, there could be a close relationship in their results. One reason for this is that their mandates generally cover the same geographic area (Atlantic to the Urals). Also, the majority of participants in the larger conference (35 nations) will be in the smaller one (23 nations). For many of those countries participating in both sets of negotiations, national representation is likely to be provided through a single Ambassador supported by a combined mission.

Experience gained in the INF process will be instructive in at least two areas associated with conventional stability and confidence-building negotiations. First, form a negotiating point of view, the INF agreement has recognized the necessity of significant asymmetric reductions to reach an agreed common level. This principle will obviously have a direct application as a precedent in reaching a common lower force level from the present recognized imbalance of forces in CFE. From the standpoint of verification, INF negotiations have already established an On-Site Inspection (OSI) regime that is relatively intrusive. Combined with the inspection implementation experience now being gained and evaluated under the OSI provisions of the Stockholm Document, agreement in the upcoming negotiations on the modalities for verifying compliance should be easier to reach. The blending of experience in the bilateral (INF) and the multilateral (Stockholm Document) areas, therefore, are mutually supportive and promise to provide a practical approach, at the outset, to problems which have, at times, (in MBFR for example) been dealt with in the abstract.

There remains, however, a broader area of uncertainty in terms of what must be verified and the means to be employed. While it is generally true that the technology exists to meet most verification problems, the definition of these problems (distinguishing a truck from a tank, for example) and the best mix of technology (space-based vis-à-vis airborne, for example) must be addressed in a coherent fashion. Recognition of the advantages of a *dedicated* verification system and its products, as compared to the use of already over-committed intelligence systems (and the concomitant necessity to degrade their product in some manner), are relevant.

18.3 Verification Methods

One basic assumption which can be made is that overhead surveillance, in some form, will play a central role in the verification of multilateral arms control agreements, as it does for USA/USSR bilateral agreements. For that reason, the means by which overhead surveillance can be carried out - airborne and/or space-based sensors - is important. Of course, overhead reconnaissance has limitations regardless of whether airborne or space-based sensors or a combination of the two are used. These limitations, and the use of additional verification means to overcome them, are evident once a practical verification system is contemplated. The application of cooperative measures and the acceptance of collateral constraints could play an exceptionally important role in increasing the effectiveness of methods used in the verification of a conventional armed forces agreement in Europe.

18.3.1 Space-Based Sensors

For some time, space-based sensors have been viewed as the tools primary for the use of superpowers in gathering data for both intelligence and verification applications. These assets were developed to gather data in areas of the world not readily accessible to airborne reconnaissance. An arrangement amounting to "open skies" for satellites has evolved between the superpowers. Provisions for satellites, although not mentioned by name but referred to under the euphemism "national technical means", have been made in major bilateral superpower arms control treaties (SALT I, II). Of course, NTM of the major powers include more than just space-based assets.

One of the main reasons for using satellites as reconnaissance platforms, then, has been that their use has been legitimized for arms control applications. In addition, spaceborne systems retain technical advantages over other means of reconnaissance when large areas of coverage are involved. They are likely to constitute the core system in a verification model.

A satellite, because of its high speed and altitude, can cover ground very quickly. A typical Synthetic Aperture Radar (SAR) satellite would be able to image the ground below it at a rate of over 300 km^2 every second at a resolution approaching seven metres. In comparison, a turbo-propelled airborne radar imaging system would cover less than five km^2 every second at a similar resolution.

The space-based sensor can, therefore, map out large areas very quickly, and provide relatively unblurred "snapshots" of events in that area. For example, the 8 million km^2 of Europe between the Atlantic and the Ural mountains, could be mapped completely every eight days with two SAR satellites.

There are several problems associated with the use of space-based sensors, however. The first is that they are extremely expensive. For example, the cost of a SAR imaging mission, including the satellite control facility and the data processing elements, was estimated at $725 million over five years.

Another disadvantage is that a satellite such as that which might be used in a multilateral verification system would take up to three or four days to provide access to a site of interest with high resolution optical sensors. Weather conditions may prevent clear viewing at any particular time, of course. Since the satellite cannot hover over the area of interest and wait for the weather to change, viewing opportunity could mean a seven or eight day delay between the time a sensor pass is requested to the time it occurs. Satellite operations, therefore, are not very flexible.

Finally, until advanced recoverable launch systems are available, once a satellite is launched, opportunities to service it, in case of malfunctions, are infrequent and extremely expensive. It may take in excess of $100 million and one or two years to plan and execute a satellite repair. The possible loss of the verification satellite (if there is only one or a small number of satellites) is likely to be more damaging than the loss of one of a small fleet of aircraft.

18.3.2 Airborne Sensors

Airborne sensors have been used almost since the invention of flight. Reconnaissance balloons equipped initially with cameras, and then also with film-developing equipment, were used in the Civil War, for example. By the time of the First World War, overhead photography from aircraft was an accepted reconnaissance technique.

Given the extensive development of aerial photography techniques, there are few missions that cannot be performed using airborne sensors. However, because of their relatively slow speed, aircraft cannot practically map out the entire European area of interest in as short a time as can satellites. It might take 200 nine-hour missions to map the estimated 8 million km^2 in that region. At two missions per day per airplane, over a dozen airplanes and twenty-five crews would be needed to do this job in eight days, given that rate of coverage.

Although they could not likely be used in the overall mapping mode, airborne sensors could well serve as a supplement to space-based sensors, giving the flexible, quick response, close-look capability, that current non-superpower satellites cannot provide.

Of course, airborne sensors must fly into a state's airspace in order to carry out their mission. In the past, this intrusion was declared to be unacceptable by East Bloc states. However, movement on the question in the Stockholm talks and limited acceptance of the use of aircraft (albeit by states in their own airspace) for verification in the Stockholm Document, open the possibility that an acceptable formula may be developed for the use of airborne sensors in multilateral treaty verification.

18.4 Coverage Problem

If a Synthetic Aperture Radar (SAR) were to image the entire continent of Europe everyday, the resultant food of information would prove almost impossible to analyse. At five metres resolution, for example, approximately 10,000 images would be produced per day. At 0.5 metres resolution that number could go to one million images per day. Clearly, however, some parts of the continent are more important than others for verification purposes. From the point of view of data manageability alone, then, it makes sense to partition the large treaty area into smaller segments.

Another practical reason to partition a large treaty area is that it may be more difficult to get agreement for intrusive verification measures with regard to some geographic parts than with regard to others. For example, some areas could be so militarily insignificant that it would be easy to agree to demilitarize them in the context of a conventional arms limitation treaty. Equally, an area could be of such strategic or tactical significance that it would be highly desirable to demilitarize it, though such an agreement may be very difficult to reach. Verification in these latter areas might be accomplished through random or periodic on-site inspections. Each area will have its own characteristics. Only some of those areas might require frequent synoptic overhead surveillance, with the attendant data storage, processing and interpretation load.

A treaty which demanded that the entire region under consideration be treated as a single block, and which ignored these non-homogeneities, could well run into difficulties when verification provisions are being drawn up. In effect, the stringent verification provisions appropriate in one area may be excessive or even counter-productive in another. If the treaty is to apply to a territory as large as the European continent, then it may be exceedingly expensive to apply the same treaty and verification provisions uniformly to the entire region. It would, in any case, not be easy to ensure a uniform and high quality of data analysis and interpretation.

18.4.1 Types of Areas

If the argument regarding non-homogenity is accepted, then one could define four types of geographic areas relative to verification:

1. Exclusion Areas: These are areas in which it would be agreed not to station troops or weapons. Such areas might be of no military significance, in which case they would not be relevant to the military balance. Alternatively, they could be of great military relevance, and agreeing to declare them Exclusion Areas could be seen as a way to enhance stability, security, and confidence.
2. Transit Areas: These are areas through which troops and weapons are allowed to move, but in which they may not be stationed. Presumably, movements would be governed by notification requirements.
3. Watch Areas: These are areas in which it is agreed to station and/or exercise troops and weapons. The number and type of troops and weapons would be governed by treaty. These areas could be permanent (fixed), or temporary - the size, longevity, and positioning to be set by treaty provisions. For example, declaration of exercise areas for large scale military exercises could be announced in yearly calendars.
4. No-Watch Areas: These are areas within which it is either impossible or undesirable to reach agreement on controls or limits respecting military activity. As a result, treaty provisions deal at best with troops and weapons which enter and leave these areas, not with that they do once inside.

Other areas may have to be defined, of course, to deal with military storage or pre-positioning areas, crisis and/or disaster situations, etc. These are not considered here to keep the argument relatively simple.

The utility of dividing the overall treaty region into areas such as those described above can be seen when one examines the verification methods that could be used in the different areas. The objective would be to use the methods most suited to the task in each area, rather than one set of methods throughout the treaty region.

18.4.2 Complementary Verification Methods

There are a significant number of other verification methods which have application as part of a verification package. These would include tagging,

Table 18.1. Some Strengths and Weaknesses of Three Verification Methods

Method	*Strength*	*Weakness*
Exit/Entry Point	- permanent coverage - limited physical intrusion	- may be relatively easy to circumvent - limited range of coverage - involves permanent intrusion of foreign nationals
On-Site Inspection	- provides highly detailed information on specific targets - can see inside buildings - short analysis time - flexibility of human presence	- highly intrusive - limited area of operation - limited duration
Overhead Reconnaissance (space)	- physically unintrusive panoramic view (snap-shot) - good for counting large numbers of items - experience	- less detailed information - large data volumes for large areas - long analysis time - intermittent coverage
Overhead Reconnaissance (air)	- physically less intrusive than ground-based methods - panoramic view - good for counting - better detail than space-based system - less expensive than space-based system	- vulnerable - operates in national air space - intermittent coverage - medium analysis time

control posts, exit/entry points, on-site inspections, and so on. For simplicity, only the latter two of these have been chosen for comparative purposes in Table 18.1.

In Table 18.1 the strengths and weaknesses of these two methods are compiled along with those for the two forms of overhead surveillance. For example, it would be difficult for an on-site inspection team to count how many vehicles participated in a large military exercise. Using overhead surveillance, though, such a count could be made relatively easily. On the other hand, an overhead sensor could not see into a building or detect decoys easily. An on-site inspection team could.

Table 18.2. Areas and Associated Verification Methods

	Verification mechanism			
Area	Exit/Entry Posts	On-Site Inspection	Overhead Surveillance Spot-Check	Overhead Surveillance Comprehensive
Exclusion		X	X	
Transit		X	X	
Watch	X	X		X
No-Watch	X			

Combined with the concept of designating non-homogeneous areas, the verification method most suited to each area can be identified. Table 18.2 shows the four types of areas and the three verification methods, and how they might be grouped. In this way, comprehensive analysis and counting of vehicles and equipment is used only in areas where there are troops and weapons. The verification system does not spend valuable time and resources vigorously examining areas in which it can be agreed not to put troops and weapons.

To cover exclusion and transit areas, random spot-checking is sufficient. If a tank is spotted in a transit zone, for example, then it constitutes a violation unless it is associated with a declared movement from one watch area to another. In such cases, the verification system can very quickly make the judgement as to whether or not a violation has occurred right on the spot.

18.4.3 Treaty Effectiveness

It may be argued that if the areas can be defined carefully, then a treaty which governs not only the *number* of troops and weapons but also their *disposition* and *movement* (the latter through notification) provides more security than one which governs only net numbers. One can also argue that a tailored system of defined areas minimizes the intrusiveness involved in verification and the opportunities for collateral data gathering. Also, watch areas can be defined so that the concentration of forces they contain does not unduly threaten attack while allowing for adequate defensive deployments.

18.5 Conclusions

The experience already gained in the verification of arms control agreements, both bilateral and multilateral, constitutes a valuable resource from which to draw in developing an effective verification model for conventional arms reductions in Europe. The combination of overhead surveillance, unattended sensors in-situ, control posts at exit/entry points and liaison teams can provide a strong verification package. Supplemented by the application of cooperative measures and collateral constraints, verification of a conventional arms reduction agreement in Europe is clearly possible.

19. Aspects of the Verification of Conventional Arms Control Measures in Europe

Klaus Jacob[1]

19.1 Introduction

In this paper I shall give an overview about some of our principal reflections about verification measures and means.[2] These reflections have always been linked to the aim of *easing verification itself* and *its acceptance* by all sides concerned. I shall then present, as a part of our studies, two different approaches of verification means; a conceptional one, which we called the "Envelope Scheme", and a technical one called "Tags".

At present, everybody is looking into the CFE Mandate and nobody knows what the subsequent negotiations might end up with. But there is no doubt that an efficient verification regime will be included.

19.2 What is the Real Problem of Verification?

Many articles about arms control end up with the deadlock that "verification remains the crucial problem...". Neither reasons nor an analysis are normally given for this pessimistic pronouncement. The following are generally perceived as crucial:

- detailed data exchange on forces, armament, military locations and armament industries;
- On-Site Inspections to verify compliance with the data exchanged (sometimes called "legal spying");
- intelligence gathering, as an inevitable co-product of On-Site Inspection and data exchange about status, structure, organization and composition of forces, as well as the infrastructure and logistics which the forces are embedded in.

Less crucial seems to be: the establishment of the verification machinery, manning, languages, equipment costs, etc.

Our team has studied a number of means and measures under the aspects of reducing the alleged "cruciality" of verification. Consequently, we have looked for specific verification means and measures of low intrusiveness, their easy handling, but effective in monitoring militarily significant non-compliance with an agreement. We have come to the conclusion that restrictions on troop levels are difficult to verify, and scarcely relevant in view of the easy circumvention and obvious reinforcement possibilities.

Reductions by complete formations with their organic equipment, such as divisions, are better verifiable through On-Site Inspections aided by national means of verification. The problem is that one needs to monitor the formations which remain. An intrusive data exchange is necessary, including the structure, strengths and armament of these remaining forces. It seems, therefore, to us more profitable to focus on reductions of military significant armaments essential for surprise attack and large scale offensive action, such as tanks, heavy artillery, armoured personnel carriers, combat helicopters and perhaps also fixed-wing combat aircraft. Besides these arms, it could be considered if specific elements of the civilian infrastructure should also come under control, such as railway bridges, flat-bottomed waggons designed for transportation of tanks, armoured personnel carriers, etc., being essential for reinforcement, concentration and large scale movements. Thus the task is to account for the number of such arms - or even better "mobile combat devices" -, which we called Treaty-Limited Items (TLIs), between the Atlantic and the Urals or in a sizable part of this area. It is for the verification of levels of such individual TLIs that we have developed two different methods.

19.3 The Envelope Scheme

This method uses as a point of departure the form of accounting we have been using in the Western European Union for many years. On the basis of a reply to a detailed questionnaire, countries or the Blocs have to state, periodically, their holdings of TLIs as well as their geographical location. The Inspectorate samples the replies through On-Site Inspections, and if these are found to be within statistical accuracy, it is assumed that the reply as a whole is correct. Tough challenge inspection procedures including free access at any time must deter the existence of undeclared barracks, depots and other possible locations holding TLIs. That is the basis of the system.

Such a verification scheme is feasible, if routine On-Site Inspections in military installations are acceptable. However, it must be realized that the scheme implies a detailed data exchange on the disposition of all TLIs. Military authorities of both sides might be reluctant to furnish all these data, which would expose the status of their forces to permanent close scrutiny. To ease the problem, we developed an on-site inspection scheme which implies divulging these delicate data only for the few places to be inspected, say 10% of the total.

We first introduce the concept of a *TLI-location*. A TLI-location is a barrack, or depot, holding TLIs, normally surrounded by a fence or something similar. On each TLI-location a designated person would be responsible for the accountancy of all TLIs present, independent of the unit or even the nation to which the TLI belongs. Once a year, the person transmits the figures of the TLI-holdings to the relevant national office. For simplicity sake, let us assume that only NATO and Warsaw Pact countries are involved, and that common ceilings on TLIs are agreed without any breakdown into national contingents. The NATO and WP offices thus receive from all their TLI-locations a breakdown of TLIs in the area of application of the agreement.

The alliances involved then exchange the following information, after being split into three separate parts: a *Location List*, a *TLI List* and a *Set of Sealed Envelopes*.

- The *Location List* reports all TLI-locations indicating only their latitude, longitude and extension. The holdings of these locations are not indicated. This allows challenge inspections to be mounted at places where it is believed that TLIs may also exist, prompted by information obtained from national means of verification, such as observation satellites.
- The *TLI List* to be transmitted would contain the number and type of TLIs at each TLI-location. However, the identity of any TLI-location would not be revealed in this list, identified by a code number set by the reporting side. This list would allow the Inspectorate to check that the overall number of TLIs was in accordance with the agreement.
- The *Set of Sealed Envelopes*, each envelope containing merely the holdings of TLIs of each single TLI-location. The envelopes would be stored by the Inspectorate under a double key system. Thus, any clandestine access to the enveloped information would not be possible for the inspecting side, while the inspected side would be prevented from making changes in the course of the year. Each envelope would only bear its national code number correlated with that given on the TLI list. Envelopes would be drawn randomly from the double key safe in the course of the year in such a way that neither of the sides would know which TLI-location had been selected until the inspection started, thus maintaining the element of surprise for both of them. When opened, the information contained in the envelope, including its exact geographical location, would be checked against the corresponding information received from the Location List, the TLI list and the inspected place itself.

Subsequently, during the inspection, the information contained in the envelope would be verified on-site, taking into accout changes which have taken place in TLI-holdings since the reporting-date. The person, responsible for the accounting on the spot, would be required to explain all changes. Just before and during inspections, observations from satellites, or - even better if agreed - from aircraft, could assist in making sure that excess TLIs were not being concealed. We prefered aircraft for various reasons. They are already in European hand, thus lessening dependence from the superpowers. Aircraft are more flexible in their flight path, can fly below the clouds and are cheaper than satellites.

In this brief description of a part of a much wider issue, I have not discussed the important problem of transport of inspectors, what exactly they would be allowed to do, etc. What I tried to explain is the advantage of the scheme, i.e. that detailed information on TLI-holdings would be disclosed only for those TLI-locations which are to be inspected, while at the same time completely random sampling is possible to verify the overall agreed levels of TLIs. Of course, during inspections detailed information is divulged to the other side about a small percentage of the TLI-locations, but any state is free to change the situation at the end of the control period, when all unopened envelopes would be either destroyed or returned to the reporting side. This scheme does not unveil the overall structure and organization of the forces. It limits the close look to a few randomly chosen places in order to verify the credibility of the reported figures and to avoid intelligence gathering as far as possible.

19.4 Tags

A completely different scheme uses "Tags", devices which would be securely affixed to TLIs, and which could not be replicated by the inspected side. The concept is as follows. Assuming that all TLIs in the area from the Atlantic to the Urals were tagged, and the quantity of tags supplied corresponded to the number of TLIs allowed in the area, then, all the inspectors would have to do, would be to verify that all TLIs they encounter were indeed properly tagged. Inspectors would be interested neither in counting TLIs nor in identifying the unit they belong to, thus largely avoiding accusations of intelligence gathering. Any TLI found without a valid tag would be an indisputable violation of the agreement, an important advantage since, in other schemes, it would often be difficult to "prove" any violation. On the other hand, if a country or alliance faithfully obeyed the rules, and untagged TLIs had not been found over the course of years, it would be less vulnerable to politically-inspired accusations of non-compliance.

An inspection scheme might, for example, take the following form. The Area of Application of the agreement could be split into a number of so-called "Inspection Area", say 100 for NATO and WP each. Both sides would be required to declare the locations where TLIs would be present, including manoeuvres. A number of Areas would be chosen for inspection each year, either randomly or on the basis of information obtained through national means of verification. Teams of inspectors would fly, to create the largest element of surprise as possible, to the most convenient airfield, and disperse to the different barracks, depots, manoeuvres and suspected locations in the Inspection Area, to check whether all TLIs they encountered were properly tagged. In addition, it might be envisaged that Exit-Entry Points would be established around the Area of Application through which all TLIs would have to be channelled when they enter or leave the Area, and tags could be attached, removed or checked at those points. Tags could also be checked at production and destruction plants. Different tags could be used if Europe were subdivided into specific Areas of Application.

Survivable and serviceable tags have to be developed, and there is some further interesting work to be undertaken by research institutes. On the requirements of a tag more details are described in our report.[3]

19.5 Conclusion

If On-Site Inspections in barracks and depots were allowed, schemes could be developed for verifying conventional arms control measures in Europe to a satisfactory and acceptable level of confidence. Much work remains to be done still technically, to develop verification equipment for perimeter controls, tags etc., and full scale exercises would have to be staged to test verification approaches. And last but not least, we should consider a verification report broadcasted worldwide as a regular and substantial part of the daily news.

19.6 Notes and References

1 The views presented in this paper do not represent official statements of WEU, and the WEU Member Countries are in no way responsible for its contents

2 For an overview of the activities which the arms control agency of the WEU performed in the field of conventional armaments, see: Past Experiences of Verifying Restrictions on Conventional Forces and Armaments, AG I(88) D/8, Paris: Western European Union, March 1988

3 Requirements of a Tag for Conventional Arms Control, AG I(88) D/10, Paris: Western European Union, 15 July 1988

20. The Verification of Conventional Disarmament Treaties by Remote Sensing

Roland Reimers

20.1 Introduction

The measures of a verification regime for conventional disarmament in Europe will comprise on-site inspection, short-range sensing, and remote sensing. It is reasonable to assume, that they will be used in a mutually supplementing manner and cross-checking each other. The scope of this paper is to present a brief survey of some technical aspects of remote sensing and to highlight some recent developments. This survey is restricted to those technologies which can be described by *imaging* with *electromagnetic* radiation.

The different technologies corresponding to this definition can be distinguished with respect to their spectral bands (wavelength λ) according to their source of illumination:

1. $0.5\ \mu m \leq \lambda \leq 2\ \mu m$: Visible light and near infrared imaging. Source of illumination is the sun.
2. $3\ \mu m \leq \lambda \leq 5\ \mu m$ and $8\ \mu m \leq \lambda \leq 12\ \mu m$: Thermal infrared imaging. Source of illumination is the thermal radiation of the objects themselves.
3. $3\ cm \leq \lambda \leq 30\ cm$: Imaging radar; for the verification purpose discussed here, a special kind: Synthetic Aperture Radar (SAR). Source of illumination is the radar transmitter.

20.2 Visible Light and Near Infrared Imaging

This technology is confined to *no obscuration* and *daylight*. It is the technology traditionally used in military and civil applications of remote sensing. It exhibits the best spatial resolutions of all remote sensing technologies: down to 10 cm ground pixel size are possible.[1] A remote sensing instrument consists of an optical part and a sensor.

20.2.1 Optics and Field of View

The optics consist of lenses or mirrors. An important constraint on the resolving power of the optics is the entrance aperture, i.e. the diameter of the entrance lens or mirror. The smaller it is, the poorer the resolution. The other optical parameter influencing the resolving power, is the focal length: the longer the focal length, the higher the magnification and the better the ground resolution. Both conditions favour large lenses with long focal lengths in order to get a high resolving power. - Fortunately, there has been a major breakthrough in lens fabrication during the last 10 years, from hand polishing of high performance lenses and mirrors to numerical controlled diamond turning by tooling machines. The optical profile is controlled by laser interferometry. Manufacturing optics in this way takes days instead of months or years as required by the old method.[2]

But just focusing on the high-resolving power optics yields a distorted view of the verification task: in order to cover the whole European area it is useless to do this with such optics only. Just to present an example: the Kodak CA-990 camera (Table 20.1) has a focal length of 2.8 m. From an altitude of 1 km it would see a strip of 3.1 m width, from 25 km altitude (reconnaissance planes) of 79 m width.

If only the Kodak CA-990 was used, the amount of data to be processed would exceed any asset available in the East and the West. It is necessary therefore to use a wide-angle camera with a comparatively worse resolving power too. The optics of this type of camera is much less sophisticated than the optics of a close-look camera.[3] Therefore, the close-look camera is reserved for spots of special interest and the wide-angle camera for the overview.

20.2.2 Sensors

Traditionally film was the medium for recording the image. This has changed during the last few years: Now focal plane arrays, primarily read out by charge-coupled device circuitry and therefore often refered to as CCDs, are mature enough to be used.[4] A CCD is a linear or rectangular array of photosensitive semiconductor pixels. The pixel size of the CCD is a limit for the resolving power of the device too. Presently pixel sizes of 15 μm have been reported and a miniaturization down to 5 μm should be possible.[5] 15 μm is already good enough for most verification purposes, particularly on a cooperative basis. Beside the structure of data obtained by a CCD, ready for computer processing, CCDs are 10 to 100 times more sensitive than film, enabling to extend the remote sensing to times of dawn.[6] But it is not possible to distinguish spectral bands with one CCD alone. Multispectral sensing is possible by parallel use of different CCDs, but only at the expense of spatial resolving power. Very good experiences have been gained with such a multispectral sensing in the Landsat and Spot satellites.[7]

The CA-990 camera from Kodak, mentioned above, is suitable for a close-look verification task in Europe. Its technical data are displayed in Table 20.1. Basically it is a close-look aerial reconnaissance camera. It can

Table 20.1. Kodak CA-990 Aerial Camera[8]

Construction	scanning mirror, primary mirror
Focal length	f = 2.8 m
Aperture	D = 50 cm
F-number	f/5.6
Detector	push-broom CCD with 64*800 pixels, pixel size = 11 μm
Data link	20 channels, 8 bit/pixel
Costs	for 2 cameras, ground station, 2 laser recorders: $50 million
Performance	
46 km	counting of sleepers
140 km	precise identification of tanks
280 km	general identification of tanks and aircraft

be installed in a great variety of aircraft. It seems to have been designed not exclusively for US military use but also for export. Oblique imaging is possible according to the figures illustrating its resolving power, although it should be noted that these numbers look a bit strange if it is taken into account, that the CA-990 is designed for aerial use only. When imaging from aircraft, such distances occur only if a very oblique angle is chosen. But this would mean that very far distances within the lower dense layers of the atmosphere have to be passed by the light rays. They would be so much distorted and attenuated, that no conceivable optics would be capable of compensating for these deficiencies. Therefore, these performance figures make sense only if imaging from a satellite is thought of. In this case only the lowest 12 km of the atmosphere have to be passed because the angle is more or less perpendicular.

20.3 Sensing with Thermal Infrared

Since thermal infrared (IR) is emitted by any object according to its temperature, it can be sensed day and night. There are already many military applications using thermal IR for night vision. Furthermore, it can penetrate haze better than visible light. The spatial resolution that can be achieved is some 10 times worse than for visible light, approximately 1 m, because the longer wavelength causes a bigger diffraction by the entrance lens of the thermal IR optical device.[9]

In principle, a thermal IR sensing instrument consists of optics and a sensor like a visible light camera. It yields a heat image, similar to a black-and-white image, where warm spots, for example chimneys or motors, appear bright. But other objects with room temperature can also be distinguished, due to their different emittance of thermal radiation. Figure 20.1

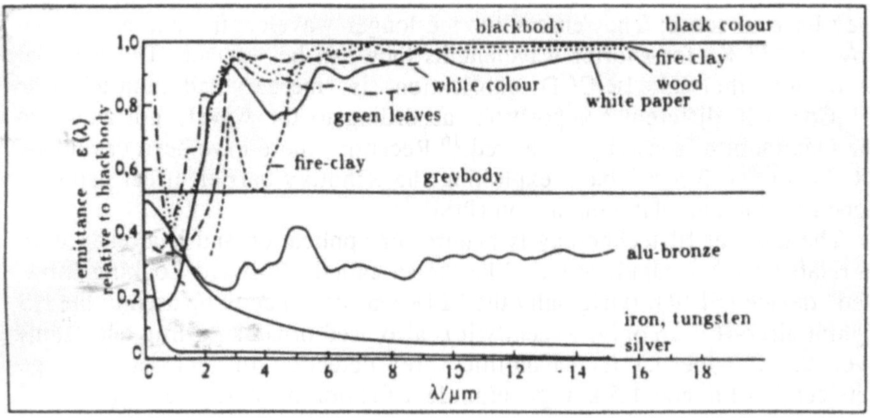

Fig. 20.1. Thermal emittances of several materials to black body radiation[10]

shows how emittances of different materials differ from each other. The emittances are given in % of the black body radiation emitted at the respective wavelength. Whereas organic substances, such as leaves, paper and wood exhibit a high emittance, metals have a very low one.

The optics of a thermal IR camera consists of lenses and/or mirrors.[11] The mirrors are made of the same material as those for visible light: fused silica.[12] Unlike the lenses for visible light, the thermal IR lenses are made of germanium which is opaque in the visible band of the electromagnetic spectrum and has the disadvantage of very high thermal expansion coefficient.[13] Therefore, some compensation, which could prove to be very sophisticated, has to be made if used as lens material.[14] For the manufacturing of thermal IR lenses and mirrors the same holds as for visible light optics.[15]

As for visible light, there are CCDs available for thermal IR though they have to meet higher requirements. The intensity of the thermal radiation emitted by a body is much lower than that of reflected sunlight and decreases with the fourth power of temperature. Two thermal IR spectral bands are distinguished according to the two atmospheric windows of transparency. The short wavelengths correspond to the maximum of radiation emitted by bodies with temperatures above 600 °C. Bodies with room temperature, about 20 °C, have their radiation maximum in the longer wavelength band. The shorter wavelength band is used for the sensors of target-finding weapons, because jet engines or hot motors appear as bright spots in this band. More interesting for verification purposes is the longer wavelength band, because primarily one has to deal with objects of room temperature. Since the vast majority of human activities is accompanied by a rise of temperature, thermal IR is a good indicator of any kind of ongoing activities. All thermal IR sensors have to be cooled well below the temperatures they have to sense: The sensors for the longer wavelength band are cooled down to 77 K, for shorter wavelengths they can be warmer. The material used for thermal IR-CCD fabrication is mercury-cadmium-telluride (HgCdTe) in different compositions according to the wavelength at which the maximum of sensivity is desired.[16] Recently, there have been developed CCDs for the 3-5 μm band exploiting the Schottky-barrier effect using p-doped silicon and platinum-silicon (PtSi).[17]

The thermal IR technology is mature for application and is already used in related military applications, like the so called "Forward Looking Infra-Red" device (FLIR). (Originally the FLIR was designed to be used in the US fighter aircraft F-16 only. Recently it is also used in US anti-drug operations over the Carribean.) Its capabilities are: detection of tanks at a range between 3.4 km and 4.5 km; general identification of tanks at a range of 2.2 km.

20.4 Synthetic Aperture Radar

Synthetic Aperture Radar (SAR) is almost completely independent of environmental conditions: it is able to obtain data day and night, during storm and rain. Since it carries its own source of illumination, the radiation can be selected appropriately to the terrain being sensed: wavelength, polarization and coherence can be altered. SAR can penetrate dry soil and vegetation to a depth of several metres, which is a useful property if camouflaged military installations are looked for. Its spatial resolving power is similar to that of thermal IR cameras, of the order of magnitude of 1 m.[18]

Since a SAR needs its own source of illumination, any SAR platform has to carry an appropriate power supply on board. This is no obstacle to aircraft with comparatively powerful jet engines. But it poses severe difficulties to satellites. Their power supply usually relies on solar panels. The electrical energy per unit area provided by today's solar panels is far too small to operate an SAR permanently. Civil SAR satellites, like the planned ERS-1, which rely on solar panels, can use their radar for only 10 minutes per orbit, which is 11% of the time. Military SAR satellites use nuclear reactors to avoid the temporary shut-down due to the lack of energy.[19]

SAR is a derivative of normal radar, exploiting somewhat different physical mechanisms. Because of the comparatively long wavelength of radar radiation, antennas of a size of several km would be necessary to gain a resolution appropriate to resolve objects on the ground. SAR circumvents this obstacle by exploiting the forward motion of the platform to synthesize an artificial ultralong antenna.

The forward motion causes a Doppler shift of the responses, depending on the position where they have been reflected from. By recording the Doppler shift together with the time elapsed between emission of the signal and reception of the response it is possible to reconstruct with extensive numerical calculations all responses reflected from one position. In this way an image can be constructed from the recorded echoes. It is important to note, that the data recorded immediately, form an interference pattern not an image.[20] Therefore, unlike other methods, a separate step for image formation is necessary. This step is so consuming in computation time, that it could not be performed in real time. With the evolution of computer technology this may become possible in the future. Data processing rates of the order 10 billion operations per second are necessary to achieve this goal.[21]

It is doubtful whether it is really desirable to have an SAR for verification, operating in *real time*. Such a system would be very useful for military targeting, battle management and similar tasks,[22] contradicting any requirement for verification. But if such a misuse could be excluded it would be a very valuable tool for verification.

20.5 Airborne or Space-Based Remote Sensing?

The next question related to the verification of arms control agreements is: what could serve as an adequate platform for the remote sensing devices, aircraft or satellite? The answer to this question depends on several factors:

1. *The size of the area to be surveyed:*
   My considerations are restricted to possible conventional disarmament measures in Europe. Verification would be necessary in the whole area between the Atlantic and the Urals. The central areas would have to be surveyed more intensely because they are closest to the line of confrontation where any violation poses the greatest risks to the national security of the states concerned. Compared to the land mass of the world Europe is quite a small area. Therefore, it can be expected that regional means are suited best for a verification task in Europe alone, resulting in a preference of aircraft. Satellites can be only a small fraction of their time over Europe which implies that it would be a waste of monitoring power if a satellite were launched only for monitoring Europe. On the other hand, Europe is not too large to be monitored from jet planes.
2. *The frequency of surveillance and the predictability of the time of overflight:*
   These conditions favour the aircraft in a similar way. Due to the periodic nature of a satellite orbit, overflights of satellites are quite predictable. The problem of predictability of satellites has been attacked, in the case of the military monitoring satellites, by providing them with a large amount of fuel enabling them to change their orbit more arbitrarily. But, on the other hand, a regular, periodic overflight is not necessary for verification. Verification is not intended to count the adversary's weapons exactly but only within a certain accuracy, to prohibit an undetected, significant break-out. From this point of view, it suffices to make unannounced overflights, thus raising considerably the probability of detecting a significant violation.

3. *The weather conditions:*
   The weather conditions also favour aircraft overflights, at least for visible light and near infrared remote sensing. Aircraft would make it possible to fly below closed ceilings of clouds and use the equipment despite adverse weather conditions. A closed ceiling of clouds would prevent effectively the use of photographic equipment from platforms above it. Particularly in winter there is often (50-60% of the time) a ceiling lower than 1 km (Fig. 20.2).
4. *The costs and the technologies available:*
   Looking at a possible verification regime for Europe, there are some peculiarities which must be taken into account:
   a) The 23 states, members of NATO or Warsaw Pact, are directly involved in any disarmament regime. Therefore, it should be demanded, that the technology for the necessary verification measures be accessible to each of them. The space-based remote sensing technology is more or less exclusively in the hands of the superpowers. Recently, a civil photographic satellite, the French/Swedish SPOT, was launched, and there are plans in France to build and launch a military, high resolution reconnaissance satellite. Thus, though there is a tendency of increased space-based remote sensing by states, other than the superpowers, their number is relatively small. It can be expected that space-based remote sensing still will remain the domain of the superpowers and that, therefore, its use for verification purposes will be more or less controlled by them. Launching vehicles are restricted to the USA, the USSR and the western European states only. The western European base, Courou, is located at the equator, and therefore not suited to launch satellites with a polar orbit. Polar orbits are neccessary for any remote sensing satellite passing over Europe. Thus, in order to have a superpower-independent launching capability it would be necessary to build a new launching centre closer to the pole.
   b) On the other hand, each country is well equipped with aircraft. The aircraft used for verification need not to be very sophisticated, any commercially available aircraft would be sufficient.
   c) Moreover, the costs of any space-based systems currently exceed those of airborne ones by a factor of approximately five to ten. To give an example: it is estimated, that the U.S. military reconnaissance satellite KH-11 costs $500 million resulting in an annual procurement of $350 million to support one satellite in orbit.[23] At least two satellites would be required for a verification monitoring task.The costs for airborne data collection, on the other hand, will hardly exceed $100 million (see Table 20.1). The large difference between these platforms can be partially accounted for by the greater technical effort that is necessary for the optics on board, to compensate for the higher altitude of satellites.

It is extremely important, to attain a status of equality among all states involved in the disarmament regime in order to avoid a possible cause for future disagreements. Furthermore, verification should not be more expensive than the weapons removed by the disarmament agreement. Thus, airborne data collection is also favoured by these aspects.

20.6 Possible Verification Tasks

What kind of measures can be envisaged in a disarmament regime? There are many suggestions of how to proceed. Possible results of successful disarmament talks can be:

- ceilings on weapons confined to certain zones (= weapon reduced zones);
- zones demilitarized of certain kinds of weapons (e.g. tank free zones);
- completely demilitarized zones;
- restrictions of military activities in certain zones;
- shut-down of military facilities presently existing.

Apart from the first, all of these disarmament measures can be verified quite precisely by remote sensing. This means that there is a good probability of detecting any violation of the above listed disarmament measures by remote sensing. As for ceilings on weapons, remote sensing is able to provide at least a qualitative account of any violation.

Remote sensing could be most useful for the monitoring of military activities, particularly manoeuvres. It is already the case, that manoeuvres must be notified in advance and that inspectors from the potential adversary are invited to observe what is going on. Remote sensing could provide an overview for these inspectors which they often have lacked in the past. It would build up confidence, because it would facilitate the assessment whether a manoeuvre is a masked surprise attack.

Anyway, to avoid a distorted impression of the verification task, it must be stressed that remote sensing will be only complementary to on-site inspection and short-range sensing. An important task, therefore, for remote sensing will be to spotlight places where there could be ambiguities about a violation. The precise answer to this question will then be given by other means.

20.7 Weather Conditions in Germany

Crucial to the choice of remote sensing technology for Europe are the weather and illumination conditions. Some values for West Germany are shown in Fig. 20.2, i.e. the percentage time of no obscuration in the four main geographic regions, for the four seasons at different times of the day. Any kind of obscuration would prevent the use of photographic remote

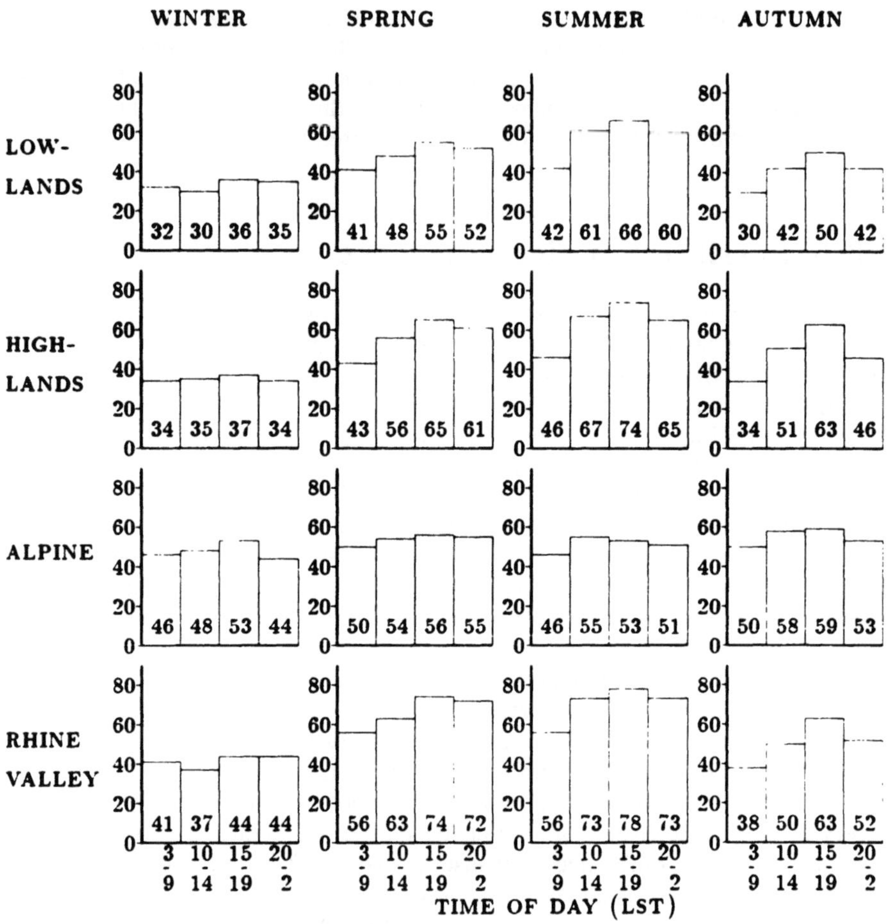

Fig. 20.2. Percentage time of no obscuration for Germany[24]

sensing. Whereas in the summer there is no obscuration for 60-80% of the daylight time, these values are 30-44% in winter, except for the alpine region which is not so important for a verification regime. Taking into account the short daylight period in winter, some 8 hours at 45° northern latitude, and remembering that areas well north of the polar circle also belong to the region to be monitored, it is concluded that in quite a large area photographic and thermal IR remote sensing from high flying aircraft or satellites is impossible during that period.

20.8 Data Processing and Data Integration

The field with the greatest deficiencies is probably data processing. A distinction can be made between three qualitatively different tasks encompassed by data processing:

1. *Pre-processing* includes the compensation of systematic, sensor-specific errors. It is applied to the collected data, in our case images, without regard to the objects imaged. It poses no difficulty and is done automatically.
2. *Restoration* and *enhancement* include all techniques for the compensation of deficiencies arising from physical, non-systematic degradations occurring during the imaging process. Restoration tries to compensate deficiencies caused by badly adjusted optics, whereas enhancement tries to accentuate features in technically "perfect" images. In the meantime there are many ready procedures for both tasks which in effect facilitate the succeding interpretation.[25]
3. *Interpretation* and *understanding* include all methods of information extraction. This can only be done by the use of *collateral information*. Therefore, an *integration* with information from other sources is required in this step. Unlike the preceding steps, this is a highly complex, non-schematic task. In traditional reconnaissance it is carried out so far by human interpreters who rely on computer assistance in the two preceding steps. For instance, the CIA currently employs in the National Photographic Interpretation Center more than 1000 interpreters.[26] In order to do the interpretation by computers new ways of computing have to be done, for example, by means of *artificial intelligence*, but this is far from being ready or applicable.[27]

20.9 Concluding Remarks

1. There are mature technologies, relying on electromagnetic radiation, appropriate for use within a verification regime.
2. All of these technologies are already in use for military purposes. One must be extremely careful not to interfere with these uses.
3. Data processing and integration raises problems so far unsolved. Presently there is neither the software nor the hardware capacity to process the amount of data to be expected from a verification regime in Europe.

20.10 Notes and References

1 A. S. Krass, How much is enough?, London: Taylor and Francis, 1985, p. 20; B. Jasani, T. Sakata (eds.), Satellites for arms control and crisis monitoring, Oxford University Press: Oxford, 1987, p. 10; J. Richelson, Technical collection and arms control, in: W. Potter (ed.), Verification and arms control, Lexington MA: Lexington Books, 1985, p. 169

2 R. E. Parks, Overview of optical manufacturing methods, in: Contemporary methods of optical fabrication, Proc. of the SPIE, vol. 306, pp. 2 ff, Bellingham WA, 1981; G. M. Sanger, Perspective on precision machining, polishing and optical requirements, in: Proc. of the SPIE, vol. 306, pp. 96 ff; G. Wright, J. B. Bryan, Proposed method of producing large optical mirrors: Single point diamond crushing followed by polishing with a small area tool, Optical Engineering, no. 9, pp. 1021 ff, 1986; T. Foley, Air Force research focuses on automatic production, Aviation Week & Space Technology, vol. 129, No. 3, pp. 74 ff, 18 July 1974; T. Foley, Phasar research to benefit SDI, intelligence satellites, Aviation Week & Space Technology, vol. 129, no. 3, pp. 72 ff, 18 July 1988; J. N. Bahcall, L. Spitzer, The Space Telescope, Scientific American, No. 7, pp. 38 ff, 1982; R. J. Ondrejka, Imaging Technologies, in: K. Tsipis, D. W. Hafemeister, P. Janeway (eds.), Arms control verification - The technologies that make it possible, McLean VA: Pergamon Brassey's International Defense Publishers, 1986, pp. 84 ff

3 R. N. Colwell (ed.), Manual of remote sensing, vol. 1, Falls Church VA: American Society of Photogrammetry, 1983, pp. 183, 276, 279

4 J. Kristian, M. M. Blouke, CCDs in astronomy, Scientific American, no. 10, pp. 66 ff, 1982; Bahcall (note 3); Colwell (note 3), p. 359

5 Bahcall (note 3)

6 Kristian (note 4)

7 P. N. Slater, Remote sensing: Optics and optical devices, Reading MA: Addison-Wesley, 1980, pp. 511 ff

8 B. D. Nordwall, Kodak team develops aerial camera with electro-optic, film capability, Aviation Week & Space Technology, pp. 101 ff, 15 August 1988; Talk given by J. Doyle, refering to Kodak information papers, at the Seminar on Independent Arms Verification and Crisis Monitoring from Space, Kiruna, Sweden, August 27-29, 1988

9 H. D. V. Böhm: Nachtsichttechnik: Einsatz infrarot-empfindlicher Wärmebildgeräte, Physik in unserer Zeit, no. 2, pp. 33 ff, 1987

10 Böhm (note 9), reprinted by permission

11 G. R. Noyes: Design and implementation of a continuous zoom FLIR optical system, in: Practical infrared optics, Proc. of the SPIE, vol. 131, 1978, pp. 24 ff; D. Rock, J. Redus et al., Falcon eye FLIR system, in: Infrared sensors and sensor fusion, Proc. of the SPIE, vol. 782, Bellingham WA, 1987, pp. 38 ff; M. F. Kessler, The Infrared Space Observatory, in: Instrumentation for optical remote sensing from space, Proc. of the SPIE, vol. 589, Bellingham WA, 1985, pp. 201 ff; M. W. Werner, J. P. Murphy et al., The SIRTF mission, in: Proc. of the SPIE, vol. 589, 1985, pp. 210 ff; J. Espiard, J. Paseri: Lightweight cold mirror and fixation, in: Proc. of the SPIE, vol. 589, 1985, pp. 187 ff

12 Kessler (note 11)

13 L. P. Mott: Energy efficient interference coatings for infrared optical systems, in: Practical infrared optics, Proc. of the SPIE, vol. 131, Bellingham WA, 1978, pp. 11 ff

14 Noyes (note 11); Rock (note 11)

15 Parks (note 2); Sanger (note 2)

16 J. C. Fraser, Infrared Surveillance and Detectors, in: Tsipis et al. (note 2), pp. 178 ff; D. A. Scribner, M. R. Kruener, C. J. Gridley: Measurement, characterisation and modeling of noise in staring infrared focal plane arrays, in: Infrared sensors and sensor fusion, Proc. of the SPIE, vol. 782, 1987, pp. 147 ff
17 A. B. Grafinger: Review of CID technology, in: Focal plane methodologies, Proc. of the SPIE, vol. 244, Bellingham WA, 1980, pp. 26 ff; R. Aguilera: 256 x 256 hybrid Schottky focal plane arrays, in: Infrared sensors and sensor fusion, Proc. of the SPIE, vol. 782, 1987, pp. 108 ff; W. F. Kosonocky, G. W. Huges: High fill factor silicide monolithic arrays, in: Proc. of the SPIE, vol. 782, 1987, pp. 114 ff
18 H. Jensen, L. C. Graham et al., Side-looking airborne radar, Scientific American, no. 10, pp. 84 ff, 1977
19 C. Elachi, T. Bicknell et al., Spaceborne SARs: Applications, techniques and technology, in: Proc. of the IEEE, no. 10, pp. 1174 ff, 1982; C. Elachi, Radar images of the earth from space; Scientific American, no. 12, pp. 46 ff, 1982; B. L. Huneycutt, Shuttle imaging radar-B/C instruments, in: Proc. of the SPIE, vol. 587, 1985, pp. 137 ff
20 Colwell (note 3), Chapters 10 and 13
21 Elachi (note 19)
22 U.S. Congress, Office of Technology Assessment, New technologies for NATO, OTA-ISC-309, Washington DC: U.S. Government Printing Office, 1987, p. 146; S. Broadbent, Joint-STARS: Force multiplier for Europe, Jane's Defence Weekly, vol. 7, pp. 729 ff, 1987
23 J. Pike, Satellite reconnaissance, Harvard International Review, no. 6, pp. 21 ff, 1988
24 Data taken from: F. E. Niles, Impact of battlefield atmospheres on E-O mm-waves, lecture presented on the 3rd annual convention on electro-optical systems and technology, Washington and Boston: AIAA, 1981
25 T. M. Lillesand, R. W. Kiefer, Remote sensing and image interpretation, New York: Wiley, 1979; R. H. C. Bates, M. J. McDonnel, Image restoration and reconstruction, Oxford: Oxford University Press, 1986
26 J. Richelson, The U.S. Intelligence Community, Cambridge MA: Ballinger, 1985
27 A. M. Din, Arms and artificial intelligence, Oxford: Oxford University Press, 1987; J. F. Judge, Strategic computing program gets high marks on midterm, Defense Electronics, no. 5, pp. 77 ff, 1986

21. Verification Techniques for Heavy Land Vehicles Using Short-Range Sensors

Jürgen Altmann

21.1 Context

21.1.1 Geographical and Numerical Limits in Europe

Changing the military confrontation in Europe in the area of conventional weapons to a status where large-scale offensive operations and surprise attack are no longer possible (as the mandate of the new Vienna Negotiations on Conventional Armed Forces in Europe states)[1], will certainly require numerical reductions on offence-capable land-mobile weapons systems, most notably the tanks and the self-propelled artillery. Limits on other heavy military land vehicles that are required for offensive operations, like mobile bridge-building gear, are useful too; armoured personnel carriers and infantry fighting vehicles may also be constrained. Most of the vehicles mentioned are tracked; this fact, as well as their size and mass, distinguish them clearly from civilian vehicles. In a comprehensive reduction scheme, it may be useful to include wheeled military vehicles, too; this could comprise wheeled troop transporters, multiple rocket launchers, missile transporter-erector-launchers, trucks carrying air defence systems. Since large-scale offensive operations require larger, and more mobile, logistics, than defence on the own territory, limits on transport trucks for personnel or supply (of munitions, fuel, food etc.) may also be introduced. Of course, the lighter and less specialized the truck categories become, the more complications arise about distinguishing them from civilian ones (in fact, in several countries it is planned to confiscate civilian trucks in times of crisis and war).[2]

Limits on military vehicles will probably be the more stringent, the shorter the distance to the inter-alliance border. This can lead to the creation of zones where different maximum values of numbers and densities for the various types of vehicles apply. For example, a forward zone of 50 to 150 km width could be completely free of tanks and artillery (light infantry vehicles may be allowed); in a second zone of 500 to 1,000 km depth the tank number may be limited to 2,000-3,000; and in the rest of Europe from the Atlantic to the Urals, the allowable number may be somewhere around 10,000-20,000 at the end of a reduction process.[3] In order to hamper formation of large offensive force concentrations, one could also introduce local movement restrictions within a reduction zone, e.g. limit the movement of tanks to their barracks and to a nearby training ground.

21.1.2 Verification Methods

Verification of numerical limits on specific mobile objects in a certain area can be done in a number of ways. In principle, one could carry out an exhaustive search and total count over the whole area repeatedly, but this is certainly unacceptable for reasons of cost and intrusiveness. A more realistic approach would use data exchanges on the local deployment numbers together with random inspections at the installations. Another random inspection method would preserve the right to move objects freely within a zone by tagging them.[4] Whereas random inspection schemes provide strong incentives for abiding by the agreement, it may nevertheless take a few years before a minor violation is detected. If this is considered a problem, one can - after an initial balance - introduce tight controls at the borders of a reduction area (and at inner points, where objects may be brought in or out, like airports). Such comprehensive control would immediately signal a violation and would thus be more effective in deterring it. Of course, permanent controls by humans on lines 1,000 km or more in length is impractical - again for reasons of cost and intrusiveness. Short-range sensors may, however, provide a cheaper and more acceptable alternative. They lend themselves naturally to continuous monitoring, they do not suffer from exhaustion or boredom. In addition, cooperatively emplaced sensor systems can be specifically tailored to collect only the information required for the agreed purpose, and this can be reliably controlled by the inspected side. Combining sensors with on-site inspection or tagging schemes is possible, of course.

This paper gives an overview over the technical possibilities of controlling the flow of treaty-limited items across the border lines of different reduction regions and areas (Fig. 21.1). One type of flow is at normal traffic lines (roads, railroads; portals of ports and airports as well as of barracks, depots, manoeuvre areas, production or destruction facilities). Here, sensors for very short distances (i.e., 0 to 20 metre) will be used. In addition, at the most important sites, human inspectors may be stationed permanently; at others, random and challenge inspections may suffice.

The other type of vehicle flow can occur between the normal traffic lines - across the perimeter of a facility, or across the border line of a reduction area (using small roads or even through cross-country or by crossing rivers). Here, sensors for greater distances (i.e., 20 m to maybe 2 km) are required.

Some properties of heavy military land vehicles deserve mentioning here: Such vehicles, in particular tracked ones, are usually transported over large distances on rail, road, on board ships or aircraft. Normally, they drive by themselves only through the manoeuvre/battle zone. The overwhelming

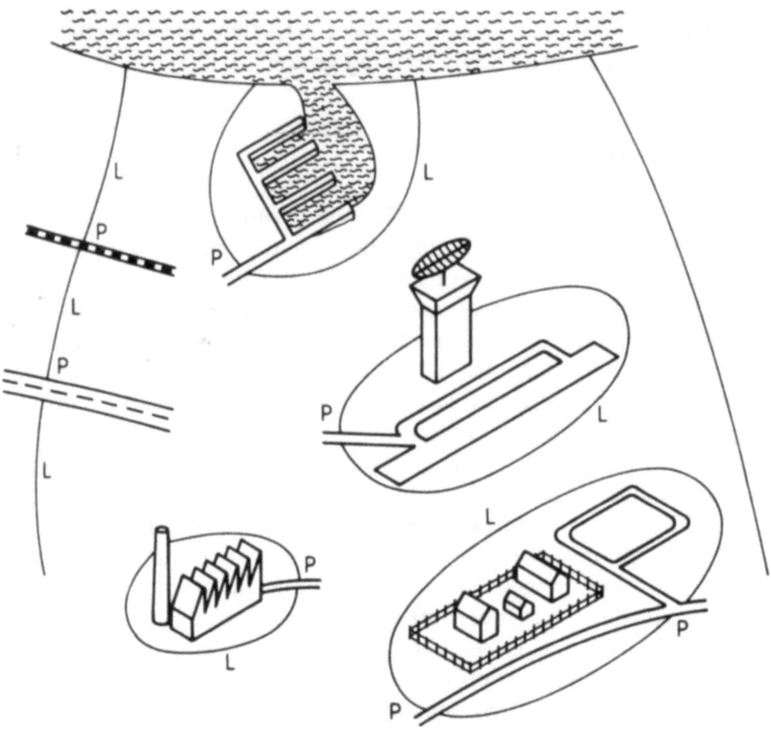

Fig. 21.1. The number of treaty-limited items in a reduction zone can change either by import or export on rail, road, via sea- or airports, or by production or destruction within the zone. A comprehensive sensor verification scheme can be constructed using two types of installations: *Point*-type controls (P) at the traffic lines and portals detect, discriminate and count treaty-limited items as they pass. *Line*-type controls (L), at some locations supplemented by physical barriers, detect or prevent illegal movement across border lines or perimeters at the sections between the point controls. Deployment of line controls along coastlines is not essential, since heavy military vehicles can only be transported by surface ships which can easily be observed by national technical means (e.g., surveillance aircraft) from international airspace.
In principle, a third category of *area* controls would be needed to detect production at secret facilities, or air transport to other locations than large airports. The latter possibility can be ruled out, however, because helicopters and vertical/short take-off and landing aircraft cannot lift masses above about 10 metric tonnes, and heavy military vehicles have masses from 15 to 60 tonnes. (Balloon or airship transport can also be excluded as impossible, or too cumbersome and too difficult to hide.) For secret production, other controls than area-wide deployment of millions of sensors are more practical. To name a few: detection of facilities by satellite or cooperative overflight; challenge inspections; tagging.
If zone-wide limits are complemented by restrictions on the mobility of treaty-limited items, sensors can also be used to delineate smaller areas, e.g. verifying that tanks do not leave the vicinity of their barracks and an adjacent training ground

portion of such vehicles is too wide or too high to fit into containers or closed railcars; thus, at monitored exit/entry points, human or automatic observation of their outer appearance will allow classification (possibly augmented by a no-cover, or transparent-cover-only, rule). Only certain classes of vehicles are capable of moving through general cross-country areas or through rivers; this will alleviate the detection and discrimination task for the sensor systems at many sections of the border line.

Separating fighting vehicles into main components (e.g. turrets and tank bodies) and secured storage over a large mutual distance may be an efficient means of introducing delay into surprise attack scenarios.[5] Securing the perimeter and the portals of such storage areas against illegal movement of parts is possible by traditional and new methods of security technology, and will not be covered here. This paper is concerned with the detection and discrimination of complete vehicles or their main components. Its focus is on Europe, but many of its considerations can be applied elsewhere as well.

21.2 Sensors for Very Short Distances

For this type of sensor system, the verification methods of the INF Treaty constitute an important precedent and role model.[6] Sensors for distances from 0 to perhaps 20 m would be used at those locations where traffic is normally occurring. Along the border line of a reduction zone, this applies to the points where roads and railroads enter the zone. At specific locations within the area - ports and airports, production and destruction facilities -, the portals would fall under this category. For completeness, point-type controls have to be augmented by line-type controls along the borderline or perimeter between adjacent points.

Two types of point-like locations can be defined. At the first one, only markedly different vehicles are allowed, e.g. a minor exit of a facility where only private cars and small vans would pass, or a minor road which has been declared out of bounds for heavy military vehicles. Here simpler equipment, which would signal an alarm on any detection of a treaty-limited item would suffice. With a right of immediate inspection in case of an alarm, this could be quite effective. At the second location type, vehicles which are identical or similar to the limited ones could legally pass. This would for example, apply to the main portal of a facility, or to the declared exit/entry points at the border of a reduction zone (Fig. 21.2). Vehicles passing would have to be detected, discriminated as treaty-limited, and categorized; the inflow and outflow of the different types would be counted in order to have a continuous balance of the number within the zone.

Different physical measures can be used. The presence of an object can be detected by ultrasound or infrared intrusion detectors. The size (length, height) can be measured by arrays of light beam interruption devices. Cameras can provide more detail, but during night-time, of course, they

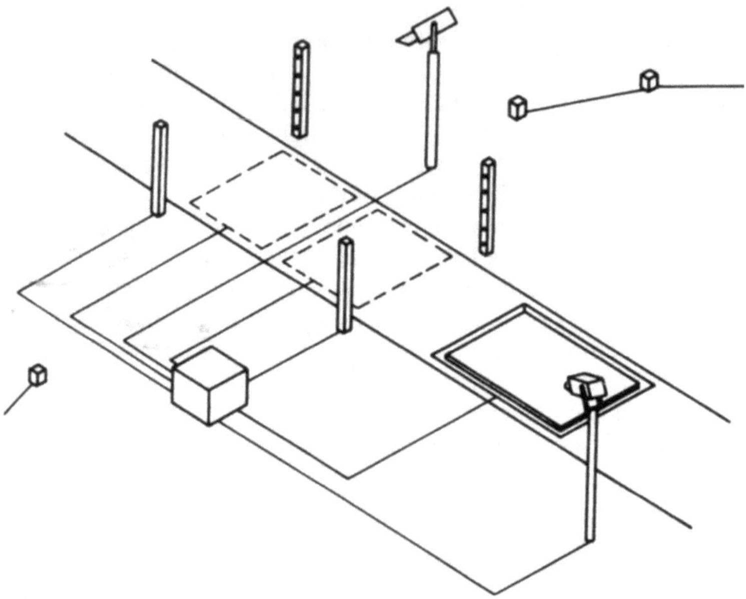

Fig. 21.2. Sketch of a sophisticated (semi- or fully) automated sensor system at a control point. Limits on size (length, height) can be checked by two linear arrays of light beam interruption devices. Induction loops detect the presence of a vehicle, its velocity and the type of propulsion (tracked or wheeled). Cameras measure size more exactly, and determine shape in addition. A weighbridge serves for measuring the vehicle mass. All sensor signals are fed into a computer, where the data are processed, reduced, stored and relayed to a verification centre. The line-type control extends on both sides of the point-type one

require illumination. The form/appearance can be roughly assessed by light beam arrays; detailed measurements require cameras. The mass of a vehicle can be determined by a weighbridge. For railcars, the weight carried by the individual axles can be measured via the deformation of the rails. The passage of a metal vehicle on a road can be detected by an induction loop in the surface. This technique allows to determine the type of propulsion (i.e., wheeled or tracked); for wheeled vehicles an estimate of the number of axles and detection of a trailer is possible, too. Table 21.1 contains a few sensor types together with typical cost figures. Signals from different sensors would be acquired and processed in a computer, which might add $10,000 (about $20,000 if camera image processing is included). In these estimates, software costs and costs of simple communication devices are included. In addition to the expenses for sensors and computer system, installation costs for cabling, housing and power have to be taken into account. Making the system weather- and tamper-proof will increase costs further. If permanent human presence is required, buildings, heating, water supply, sewage etc. have to be added.

Table 21.1. Possible sensor types for very short distances, and their approximate cost in $1,000

Infrared intrusion detector	0.5
Ultrasound intrusion detector	0.5
Linear array of light beam interruption devices	2
Induction loop with control	10
TV camera with electric zoom and tilt	8
Large vehicle weighbridge	40
Train weighing station	300

A simple sensor-computer system designed to detect any vehicle above a certain size, and capable of differentiating between wheeled and tracked vehicles, may require two arrays of light beam interruption devices and two induction loops. Taking into account a tolerance factor of 2 to 3 for increases due to redundancy and weather- and tamper-proof construction, such a location - including installation - may cost no more than $100,000. A sophisticated system capable of discriminating civilian from military vehicles, and classifying different military vehicles according to mass and form, may require expenditures in the range of $1 million; if facilities for inspectors are added, the cost would increase to some extent.

All the sensor techniques mentioned are commercially available at present, and are routinely being used today in the fields of security and traffic control. For the purpose of verification, some adaptation is required; engineering design of optimum systems - what combination of signatures should be used at which location type, developing the computer software, etc. - can begin immediately.

The total cost of such point-type control systems can be estimated, if the number of installations is known. For the forward and middle zones (up to a depth of 1,000 to 1,500 km), a number of 500 to 1000 installations may be applicable. Taking an average cost of $1 million, one arrives at a total of $500 million to $1 billion.

21.3 Sensors for Near Distances

For this class of sensors a precedent exists with the U.S. Sinai Field Mission which controlled the armistice between Egypt and Israel from 1976 to 1982.[7] Sensors for object distances from about 20 m to maybe 2 km would be used along the border line of a reduction zone, or along the perimeter of a facility, between the point controls described already. (If the perimeter is fenced, irregular checks of the integrity of the fence may suffice; otherwise, microphones - sensing attempts of cutting wire - optic fibres - embedded in

Table 21.2. Sensors which could be used for near distances (20 m to perhaps 2 km), and typical costs in $1,000. The cost is highly dependent on the required sensitivity, which has to be determined by research and development

Seismic sensor	0.1 - 5
Microphone	0.1 - 1
Underwater microphone	0.5 - 2
Infrared detector	0.5 - 5
Infrared camera	20 - 60
Magnetic field sensor	0.3 - 5

the fence, carrying light pulses - etc. may be used. Such devices will not be discussed here; the attention is on sensors which can be installed at lines where no fence exists). Of course, no transport of treaty-limited items across the line is allowed legally, and any heavy military vehicle detected would immediately constitute a violation.

The physical signatures which can be used to sense the presence and/or movement of a heavy vehicle over greater distances are somewhat different from those for very short ranges. Movement of vehicles produces ground vibration which is stronger on rough terrain, and especially with tracked propulsion. Seismic sensors are able to pick up such signals. Acoustic noise, especially from the engines, can be measured using microphones. If rivers are used as border lines, attempts to circumvent the control points by swimming or snorkeling between the bridges can be detected by underwater microphones.[8] Visible light cameras would mainly work at day and in fair weather; infrared cameras could also be used at night and in moderate haze. Another potential signature is the disturbance of the earth magnetic field by a large ferrous mass. Except for the cameras, all sensors mentioned would work under all weather conditions.[9] Table 21.2 gives sensor types and typical costs. Similarly to the point type control, a computer system would acquire and process the signals of all sensors along a given section of the line, the length of which may be a few kilometres.

For all the signatures mentioned, sensors are available commercially at present. Some are regularly used, e.g. in geophysical research. Their cost varies according to the required sensitivity and precision. Not much is known (at least not publicly) about the signatures of the different vehicle types, on different ground, with different velocities etc. The expected levels of environmental noise in the verification context are also not well known. Here, a considerable amount of research and development is needed.[10] Another topic of research should be the prospects of new sensor principles and their sensitivities, e.g. the integrated semiconductor micromechanical sensors which can be mass-produced at about $100 per copy. Object signatures, noise levels and sensitivities have to be analysed together to arrive at

Table 21.3. Rough estimate of the installation cost of a possible sensor line of 1,000 km length. The cost of cable and laying are based on statements of West German firms. The estimated cost of the data centres contains the communication equipment. Higher requirements on sensor sensitivy, on redundancy, on weather- and tamper-proof construction, can increase the total cost; this is taken into account by a tolerance factor of 2 to 3

			Total, $ m.
Seismic sensors	10 per km	10,000 * $300	3
Microphones	10 per km	10,000 * $300	3
Cable, laying	1,000 km at $20,000 each		20
Sensor housings, installation		10,000 * $1,000	10
Local computers	1 per 2 km	500 * $10,000	5
Data centers	10 total	10 * $0.5 m.	5
Total:			$44 million
Tolerance factor for more sensitive sensors, redundancy, weather- and tamper-proof construction:			2 - 3
Rough estimate for total cost for 1,000 km line:			$100 million

maximum detection distances under different conditions. Very preliminary measurements and theoretical considerations suggest that for ground vibration and sound, detection distances of several hundreds of metres may be achieved, whereas the magnetic disturbance will only be detectable - using traditional high-sensitivity equipment - over a few tens of metres (however, further technical development seems possible).

Is it economically feasible to equip the complete border line of maybe 1,000 km length between two reduction zones with sensor systems? Table 21.3 contains a rough estimate of the installation cost of such a line, where it has been assumed that one simple seismic sensor of $300, and one simple microphone of the same cost is installed every 100 metres. Including a safety factor of 2 to 3 for higher expenses, one arrives at a total cost on the order of $100 million for a 1,000 km line. Assuming a second line of 2,000 km length more to the rear, and including the $500 million to $1 billion for the point controls, installation of a complete sensor system can cost $0.8 to 1.3 billion per side. This sounds high; however, it will be distributed over a couple of years and it has to be compared with the roughly 250-fold higher expenses which e.g. NATO countries pay annually for forces and armaments in Europe, a considerable portion of which will be saved by reductions. The conclusion is that sensor verification schemes would even remain economically feasible, if their cost were higher by another considerable factor.

21.4 General Problems

When the detection distances have been determined and the sensor types to be used have been defined, several other design problems must be addressed. This concerns e.g. the communication structure which links the point and line controls to the verification centres: What kind of lines should be used (telephone network, dedicated lines, satellite links)? How much hierarchy and redundancy have to be built into the total system? How much preprocessing and decision is to be made in a decentralized way at the local data computers? Should there be a possibility for the alliance-wide verification centres to make a connection to any on-site sensor, picking up its original signal befor preprocessing? One of the arguments that has to be pondered is that the information provided may not facilitate surprise attack (e.g., coordinates of many mobile weapon systems may not be transmitted in real time). The systems have to be fail-safe, weather-proof and tamper-resistant; attempts of tampering should leave traces. In many cases, the sensors provide warning by themselves, e.g. if an attempt is made to dig up a buried seismic sensor.

Making a sensor verification system work reliably requires agreement on a set of procedures. To give a few examples: Personnel need access to sensor systems for regular service, for random inspections, in case of signaled alarm. Specific rules have to prevent saturation of sensors (e.g. earth-moving work in the vicinity of seismic sensors has to be banned, or may only be allowed after notification and in the presence of inspectors). It may also be necessary to introduce procedures to facilitate observation (e.g. treaty-limited items may only be transported uncovered across declared exit/ entry points).

21.5 Conclusion

Sensor schemes offer a prospect of reliable verification of numerical limits on different types of heavy military land vehicles in different zones. They can be tailored to the specific measurement purpose, and implemented at an acceptable cost. For a specific assessment of their properties, a considerable amount of research, development and testing is required. Certainly, the new Vienna Talks on Conventional Armed Forces in Europe will take a few years to arrive at an agreement. To some extent, this hinges on the existence of adequate cooperative verification methods. In order to promote the process of reaching agreement, these methods should be available as early as possible. On the national and alliance level, research, development and test activities could and should begin (or be strongly increased) right now. Test installations could be arranged between the blocs in a few years, before formal agreement on reductions and verification methods is reached. Over several consecutive phases, the sides could learn from the experiences made, and could modify the verification equipment and the procedures - as well as the limits themselves - appropriately. In such a way, a cooperative verification scheme can be designed and evolved cooperatively in parallel to the negotiating process.

21.6 Notes and References

1 See e.g.: Bulletin, no. 10, pp. 96-99, Bonn: Presse- und Informationsamt der Bundesregierung, 31 January 1989

2 This applies e.g. for the Federal Republic of Germany through its Emergency Powers Acts

3 The depths and numbers given serve as examples only. Zones can also be defined by countries and military districts; in order to make optimum use of the sensor concept, it is advisable to have zone boundaries coincide with geographic borders like mountain ranges or major rivers. For proposals comprising different limits in different zones, see e.g.: J. Dean, Will Negotiated Force Reductions Build Down the NATO - Warsaw Pact Confrontation?, The Washington Quarterly, vol. 11, no. 2, pp. 69-84, Spring 1988; Defensive Zonen - Stellungnahme des IFSH zur Strukturellen Nichtangriffsfähigkeit und Konventionellen Stabilität in Europa, in: E. Bahr, D. Lutz (Eds.), Gemeinsame Sicherheit, Konventionelle Stabilität, vol. III, Baden-Baden: Nomos, 1988; A. v. Müller, A. Karkoszka, A Modified Approach to Conventional Arms Control, in: A. v. Bülow, H. Funk, A. von Müller, Sicherheit für Europa, Koblenz: Bernard & Graefe, 1988; USSR Academy of Sciences, Institute of World Economy and International Relations, Disarmament and Security Yearbook 1987, Chs. 17 and 19, Moscow: Novosti, 1988. The official NATO and WTO proposals tabled in the Vienna CFE negotiations also foresee the creation of zones with different sublimits, see e.g.: Focus on Vienna, no. 13, July 1989.

4 See: K. Jacob, Aspects of the Verification of Conventional Arms Control Measures in Europe, this volume, Ch. 19

5 See e.g.: Dean (note 3)

6 Treaty between the United States of America and the Union of Socialist Republics on the Elimination of Their Intermediate-Range and Shorter-Range Missiles, 100th Congress, 2d Session, Senate, Treaty Doc. 100-11, Washington DC: U.S. Government Printing Office; see also: R. Summers, Experience with INF Treaty Verification and Prospects for Effective Verification of Strategic Arms Reductions, this volume, Ch. 2

7 See: B. S. Mandell, The Sinai Experience: Lessons in Multimethod Arms Control Verification and Risk Management, Arms Control Verification Studies, no. 3, Ottawa: Department of External Affairs, September 1987; D. Barton, The Sinai peacekeeping experience, a verification paradigm for Europe, in: World Armaments and Disarmament, SIPRI Yearbook 1985, London: Taylor & Francis, 1985, pp. 541-573

8 See also: I. Oelrich, V. Utgoff, Confidence Building with Unmanned Sensors in Europe, in: B. M. Blechman (ed.), Technology and the Limitation of International Conflict, Lanham MD: University Press of America, forthcoming

9 Note that seismic and acoustic sensors can in parallel detect low-flying aircraft and helicopters. This can be used along border lines as well as for counting take-offs and landings at air bases. See: J. Altmann, B. Gonsior, Nahsensoren für die kooperative Verifikation der Abrüstung von konventionellen Waffen, Sicherheit und Frieden, vol. 7, no. 2, pp. 77-82, 1989

10 The research project "New technical means for cooperative verification in Europe" at the Department of Physics, Ruhr-Universität Bochum, Bochum, FRG, is devoted to some of these questions. This project is financed by a grant from the Stiftung Volkswagenwerk

Part VIII

Verification of a Chemical Weapons Ban

22. Challenge Inspections in a Chemical Weapons Convention

Nikita P. Smidovich

22.1 Introduction

The problem of challenge inspections is the key problem at the Chemical Weapons (CW) negotiations and in the whole CW Convention. The overall reliability of international verification under the convention depends on how effective this very special form of inspections will be. Moreover, progress in a whole set of other articles of the Convention also depends on the earliest solution of the problem of challenge inspections. The mechanisms now being elaborated for the CW Convention may be a major contribution of these negotiations for the theory and practice of verification in general.

The idea behind the machinery of challenge inspection is to provide a means of verification over the undeclared sites (the declared sites will be subject to systematic or permanent international verification under the CW Convention). This is a highly important problem. Historically, we have paid most attention to nuclear weapons, where we have a low number of very specifically delineated facilities which were born out of state activities. As we are enlarging the scope of disarmament negotiations, we are moving to more traditional fields of the arms race. These armaments, with chemical weapons among them, are much more closely associated with conventional industry. There will be declared sites for production of weapons; but there will also be many other sites and facilities which are capable of making several kinds of products, including weapons. The challenge inspections we are negotiating now are aimed at those sites which have not been specifically declared.

22.2 Mandatory Character

In his statement at the Conference on Disarmament on August 6, 1987, the Minister for Foreign Affairs of the USSR, E. A. Shevardnadze, said that "The Soviet delegation at the negotiations on this question will proceed from the need to make legally binding the principle of mandatory challenge inspections without right of refusal".

Productive efforts are being made in order to solve the problem of challenge inspections and the negotiators are deep into details of procedures of conducting such inspections.

The mechanism of challenge inspections to be included in the CW Convention must ensure that no fact of violating the convention and the consequences of such violation could be concealed by a state, or go undetected. On this ultimate capability of challenge inspections rests its deterrent character.

Of course, challenge inspections will not be the only tool in the hands of state parties, but it will be the most powerful one. For this, the requirement for non-refusal of such a challenge is mandatory. The power of this mechanism rests on the firm knowledge of the state launching a challenge, that the suspected violator could not refuse such an inspection.

As soon as we start discussing non-refusal, we get into two problems. The first is a legal one: we have to adjust the legal system in every country to accommodate such an inspection. This applies to countries with privately owned industry as well as to those countries with a non-market economy. Secondly, it goes to the more philosophical problem of how to understand sovereignty in the present era. You cannot negotiate a non-refusal or mandatory challenge inspection if you stick to the traditional perception of sovereignty, at least as defined in many legal books, namely as a supremacy of the national law over the international law. I think that this old perception of sovereignty really is at the roots of many problems that we encounter about general acceptance of the challenge inspection procedures.

In challenge inspections, time is an important factor. It appears that no more than 48 hours should pass between the time of the submission of a challenge and the arrival of the international inspection team at the inspection site. Such a short notice is intended so that a violator could not clear the violation or conceal the facts of the violation. It is not specifically related to the time needed for such clearing or concealing; the negotiators specified 48 hours mainly because of the time needed to reach the site.

22.3 Equal Rights, Equal Application

Any state party to the convention, without exception, should have the right to make a request for an inspection. Everybody must have the same rights in this regard, as this is a fundamental right of a state-party to the convention. This might sound commonplace, but when one considers a global convention, i.e. one that encompasses not only states of "equal" military ability, especially in relation to chemical weapons, there may be a temptation to introduce certain privileges for some, and disadvantages for other, states. However, equal rights with regard to this ultimate instrument of verification in the CW Convention is indeed fundamental. I am quite sure that, if we will introduce any discrimination among member states, we will not get a global convention. Based on this, we cannot accept any filter in the mechanism of challenge inspections, such as embodied in the US proposal to establish a "fact-finding panel" with a limited number of privileged states, where only the representatives of those states can launch a full-fledged challenge, and all others have to go through that panel to get their challenge implemented. This concept does not ensure equal rights for all parties to the convention.

There should be no discrimination on the basis of the form of ownership of the locations and facilities which have to be inspected. If one starts to delineate places, which may or may not be open for inspections, one will create a certain sanctuary where the state will know that it will not get a challenge inspection. In that case, these will be the most appropriate places for a potential violator to conduct illegal activities. Thus, a request for inspection may be submitted in relation to any facility or location on the territory of a state-party, or under its jurisdiction or control, as well as in relation to those owned by any natural or legal person of a state-party wherever they are situated. This is a necessary condition to make challenge inspections global and a really effective instrument.

In this respect, the challenge procedures we are elaborating now differ from previous arms control agreements. In the Stockholm Agreement, inspections are limited to large regions and to verification of troop numbers in manoeuvres. Moreover, within these large regions there are certain areas which are excluded from verification. In the INF Agreement, although the procedures may seem similar to challenge inspection, there are only declared facilities, and the procedures apply only to those. There is a big difference between applying challenge or other inspection procedures to a declared facility - a facility prepared for an inspection -, and, on the other hand, promising to open any facility for an international inspection. On the verification side of the CW Convention, we are trying to elaborate procedures that will open every place and every facility of a state party. This may be a disadvantage for the CW negotiations, since chemical weapons are a rather small part of the military arsenals. On the other hand, I hope that it will provide a great advantage and a great contribution for other negotiations. If we can solve these problems in the CW convention, this can serve as a precedent for many other areas of disarmament.

22.4 Avoiding the Use for Other Purposes

There is one important problem which is always present at the discussions of challenge inspection - the possible disclosure of sensitive data, which can happen during such inspections. Apprehensions in this regard are of course valid, be it for military or commercial industry reasons, or in the interest of technological and intellectual property. There is an understanding that measures should be elaborated with a view to minimize this danger. A valuable idea is contained in the British proposal on challenge inspections, i.e. the possibility for the requested state to suggest alternative measures for conducting inspections in order to prove compliance with its obligations, in case of a very sensitive location or facility.[1] The possibility of using alternative measures is generally recognized by the participants in the conference. Whether the suggested alternative measures are satisfactory should be decided only by the requesting state.

It goes without saying that the time limits for agreeing on the procedure of conducting challenge inspections must be clearly defined. We must avoid any barrier that can prolong the beginning of an inspection, render it ineffective or even obscure its main purpose. At the negotiations, we try to devise appropriate specific procedures for conducting challenge inspections. This is very laborious work, in which on one hand we borrow something from the INF Agreement, on the other hand we devise completely new procedures because of the undeclared nature of the facilities being inspected. We even do not know what facilities may come under challenge inspections. We have already settled some general principles, although not final. Now the negotiations deal mainly with the procedural details.

To preclude the possibility of using challenge inspections for purposes incompatible with the task of verifying compliance with the CW Convention, or for disclosing secrets unrelated to chemical weapons, the Convention should envisage the appropriate specific procedures for conducting challenge inspections. It seems that measures need to be developed to preclude the possibility of using inspections for obtaining intelligence data. In particular, efforts should be made to ensure that the methodologies and instruments used by international inspectors in the course of inspections be strictly limited to their verification tasks. The instruments should be standard and uniform for all states parties. An inspected party should have access to the instruments for checking them.

Measures which would prevent the use of inspections for purposes incompatible with the tasks of verifying compliance with the Convention should be elaborated within the framework of - and not contrary to - the principle of mandatory inspections; such measures must not weaken this principle or make any exceptions to it.

22.5 The Position of the USSR

The Soviet position on the challenge inspections rests on the following points:

1. All states parties to the CW Convention should have equal rights and obligations in regard to both submitting a request and accomodating it.
2. A request for inspection can be submitted in relation to any facility or location.
3. A request should contain the necessary basic data (what, where, when, how).
4. The period between the time of request and the arrival of the inspectors at the inspection site should not exceed 48 hours.
5. It is important to elaborate specific measures in order to prevent the use of challenge inspection for purposes incompatible with the task of verifying compliance with the Convention.

6. The requested state may suggest alternative measures. Whether these are satisfactory shall be decided by the requesting state. The time for reaching agreement on the verification procedure should not exceed 48 hours (during the same period in which inspectors arrive at the inspection site).
7. Most important: Challenge inspections should be mandatory, without the right of a state to refuse such inspections; they should be legally binding in the CW Convention.

22.6 Notes and References

1 Conference on Disarmament, Document CD/715

23. Verification Procedures for a Chemical Weapons Treaty

Julian P. Perry Robinson

23.1 Introduction

The 1925 Geneva Protocol prohibits chemical warfare (CW) but not possession of CW weapons. The question of further CW arms control has occupied intergovernmental talks for twenty years now, since the time of the Vietnam War. An international convention seeking "for the sake of all mankind to completely exclude the possibility of the use" of the weapons is being negotiated at the 40-nation Conference on Disarmament (CD) in Geneva.[1]

The negotiation has been advancing, since 1984, through the gradual elaboration of a single set of draft treaty articles and associated annexes, this "rolling text" serving as a non-binding register of emergent consensus. The foregoing quotation is from its preamble. An interlocking structure of consultations and working parties under the auspices of the CD's Ad Hoc Committee on Chemical Weapons, whose chair changes each year in a rotation from Non-Aligned to East to West, provides the machinery of negotiation. The delegations of the Soviet Union and the United States maintain close bilateral contacts which now control, though not wholly, the pace and the direction of the overall enterprise.

The details of what has been happening in all this are not widely known, for CW commands only sporadic popular attention and the way it is regarded is much afflicted by misinformation, even disinformation. Yet in those details lie matters of not a little security significance; and the ideas and experience which they embody could benefit other areas of arms control. So the aim of this paper is to identify for wider notice some of the more singular features of the approach to verification that is being followed in the CW negotiations.[2]

23.2 The Likely Form of the Projected Chemical Weapons Treaty

It has long been agreed by the negotiating governments that the projected Chemical Weapons Convention should require the permanent elimination of all CW weapons, comprehensively defined; that it should be global in application; and that it should be properly verified by international means. Lesser goals have occasionally been mooted. There has been, for example, the notion of a quantitative agreement which would allow states parties to maintain CW forces up to an agreed ceiling; or a CW nonproliferation treaty; or regional bans; or - a long time ago now - a declaratory (i.e. unverified) agreement. But such proposals have not taken hold, save, in the case of regional ones, as possible staging posts on the way to global agreement. The residual-forces proposal is the most recent, and has been expressly repudiated by both the Soviet Union and the United States. There is widespread recognition now that a CW arms control agreement which provides for less than comprehensive CW disarmament is, under today's circumstances, very likely to prove counterproductive; so, too, a régime that is significantly less than global in its adherence, given the accessibility of CW weapons, even ones of mass destruction, to states at rather widely differing levels of economic and technological development. And the 1972 Biological Weapons Convention has shown something of the problems which a poorly verified treaty can create.

These three requirements - of comprehensiveness, universality and verifiability - are the bedrock of the currently registered consensus. They direct the present approach to verification. In practice there will have to be trade-offs between them, for they are not totally compatible. What exactly comprehensiveness of scope is to mean is, accordingly, still obscure in the rolling text, where the key part, article II, still has much space for negotiation in its array of dissenting and qualifying footnotes and bracketted alternative wordings. The general requirement, however, is clearly established there: comprehensiveness of scope is to be secured by means of a *general purpose criterion*, whereby, in effect, the treaty will outlaw possession of all chemicals except where they are intended for the listed purposes not prohibited by the treaty. The negotiators are under pressure from various quarters to narrow the scope: in some way to exempt, for example, herbicides and tear gases. The prospects for universality may be suffering in consequence. But at the same time there is also pressure in favour of breadth. This has to do, above all, with chemicals which could not themselves constitute threatening CW weapons but which might bring such weapons threateningly close. These chemicals include what the diplomats have come to call "precursors", meaning chemicals capable of ready conversion into CW agents, whether as loading for binary munitions or as plant-process production intermediates. They also include "adjuvants": those chemicals such as thickeners, stabilizers or penetrants that might be used to enhance the effects of CW agents. These are the categories of chemical - lacking, as they do, toxicity from among their necessary characteristics - which have caused the negotiators to abandon the concept of "CW agent" as determinant of scope, and to adopt instead the general purpose criterion.

Their action has this implication: if the scope of the treaty is to be cast in quite such broad language, the treaty will inevitably embrace many chemicals and associated activities which are not seriously threatening at all and which will therefore constitute elements of the treaty's scope where verification of compliance could hardly, on security grounds, be regarded as essential, even if it were feasible. Embedded in the rolling text as part of the prevailing consensus, then, is acceptance of the proposition that the scope of the treaty's verification system need not be as extensive as the scope of the treaty itself. Here is one of the more singular features of the CW negotiations.

The general purpose criterion is there, in article II, to define the "chemical weapons" that are to be subject to the proscriptions of the treaty. The proscriptions themselves, set out in article I of the rolling text, include both positive and negative obligations in order to give the desired comprehensiveness of scope. The positive obligations will require states parties to destroy their CW weapons and such factories as they may have for making them. A disarmament or *elimination régime* is thereby to be established. It is likely to affect only a minority of states parties, however, because only a minority of states today possess the weapons or the factories. Alongside it there is to be the non-armament, or *preclusion régime*, that flows from the negative obligations. This second régime will affect all states parties, or at least all of those that wish to continue benefitting from the security afforded by the treaty; for they will have to take positive steps to make sure that such potential adversaries as they or their allies may have, can see that they are indeed refraining from developing, producing or stockpiling CW weapons and everything else which article I has enjoined them not to do. They will of course be looking to the treaty's verification system to assist them in this building of confidence and assurance.

This duality of objective - seeking to establish both an elimination and a preclusion régime - may be flagged as a second remarkable feature of the CW negotiations, another direct consequence of the comprehensiveness that is sought. It can be seen at once that, as between the two types of régime, the verifiability that is desired in the overall treaty must take on two significantly different meanings. Upto a point (discussed further below), certainty of compliance can at least be contemplated in the case of elimination. But, for preclusion, verification in that strict sense would always require an unattainable proving of negatives; the best that can be hoped for is not certainty but an acceptable degree of mutual confidence in compliance. The overall verification system is to be designed, then, so as to exploit, as effectively as possible, such verifiability as may be drafted into the treaty in these two different senses - to come as near as possible to providing verification of elimination, and to build and then sustain confidence in preclusion.

Whether states judge themselves likely to be better off inside or outside the treaty, having regard to all the various costs associated with either option, will largely depend on the extent to which the system is thought capable of doing these two things.

23.3 The Verification System Now Envisaged

The basic component of that dual-purpose verification system is, as has long been agreed, to be a data-base constructed and maintained from otherwise private information declared by the states parties themselves. The treaty is to specify precisely the types of information required of each party and the modalities of its declaration. One set of declarations will comprise information about any CW capabilities to be renounced under the treaty: identities and quantities of stockpiled CW agents, particulars of production capacity for CW weapons, and the like. It will also include information about the steps being taken to eliminate those capabilities within the time periods prescribed by the treaty. A second set of declarations will comprise information about non-CW capabilities which could serve, or be confused with, CW purposes, for example, production capacity within civilian chemical industry for chemicals that could in principle be used in or for CW weapons.

The full range of uses to which that data-base is put is still under negotiation. In principle it could contain - and in practice it will probably have to include - proprietary information of a commercially sensitive nature. This means that such information may not in fact enter the data-base, whatever international law may say, unless rather strict measures can be developed for controlling access to it. Subject to such access constraints, the data-base would then serve two broad functions. It would focus the treaty's verification procedures. And by, for example, providing "ground truths", it would enhance such national technical means of verification that might be available to individual states parties - improving their abilities to discriminate between true and false signals of adversary non-compliance and to resolve ambiguities.

Little further can usefully be said here about that second function, except to emphasize its potential importance, and to observe that the relationships it would require between verification and intelligence would not be unproblematic. As for the first function, the present state of consensus is that the key parts of the declared information should be subject to international validation by routine measures of verification, including systematic international on-site inspection and, where appropriate, the use of "black box" instrumentation. To this end, an international inspectorate would be established as part of a new international agency which, in the rolling text, is called the Organization for the Prohibition of Chemical Weapons.

Operating, thus, through international validation of national declarations, the overall verification system would have two principal shortcomings. One is that not all of the key declared information would lend itself to validation through feasible routine measures. The other is vulnerability to incomplete declarations: the system could provide essentially zero assurance that stockpiles or even factories had not been deliberately omitted from the declarations and thereafter maintained clandestinely. The remedy currently being negotiated is the device of "challenge" inspection: procedures whereby a state party seeking clarification or resolution of suspicions could activate special channels, including the possibility of on-site investigation, the challenged party being under strict obligation to cooperate.

In this coordination of routine and challenge verification lies a third special feature of the CW negotiations, one that has found some reflection in the INF agreement. There are rich trade-off possibilities; for the greater the reliance that can be placed on challenge procedures, the less need to be demanded of the declarations and their systematic validation; and vice versa. Exactly where and how the balance is to be struck is still a matter of close negotiation, but it seems to be emerging that the trade-off will be rather different as between the preclusion and the elimination régimes.

Maybe, in fact, neither régime could withstand the political stresses from challenges that are at all frequent, in which case challenge procedures had better be regarded in both of them as last resort, with as much of the emphasis as possible being placed on the routine procedures. But in the preclusion régime it is the civilian chemical industry that is the primary actor; and, whether that industry be state-owned or private, such is the structure and competitiveness of the world chemical market that even quite modest requirements to provide proprietary information could be regarded as just too threatening for compliance if the information had to be provided on a regular routine basis. Hence the current idea of 'ad hoc' inspections as the primary instrument of civil-industry control: challenge inspections that are initiated, not by suspicious states parties, but by the international inspectorate itself - i.e. almost as a matter of routine.

Clearly, the long-term prospects for the well-being of the Chemical Weapons Convention, once it has entered into force, will depend on its preclusion régime: on the propensity of that régime for adapting itself as necessary over time to political and, especially, technological change, and on the degree to which the régime attracts and sustains the support and cooperation of civilian chemical industry. It has long been part of the basic consensus that new controls will have to be applied to the industry in the interests of promoting the verifiability of the preclusion obligations; and much progress has been, and is being, made on the details of these controls. This central involvement of civilian industry is the final singularity of the CW negotiations to be flagged in this paper.

The present approach to civil-industry controls may be summarized as follows. The negotiators, having chosen to follow the route of the general purpose criterion, are compelled to think primarily in terms of chemicals per se rather than plant, equipment or technology as the primary objects of control. So they have had the idea of classifying chemicals into different categories, and applying to each category a control régime of different stringency. The most "threatening" chemicals should go into the most tightly controlled category. The least threatening should remain outside the control categories altogether, although still inside the scope of the treaty itself. Table 23.1 shows the chemicals - in fact rather few in number - which the negotiators have thus far agreed should be subject to control.

Three control régimes are envisaged. Chemicals in the first of them would effectively be banned from civilian industry and commerce, whether as end products or as intermediates. They could not be banned altogether, but such supplies of them as might be needed - for pharmacological or

Table 23.1. The chemicals currently scheduled for control

Chemicals thus far listed for the *most stringent* control régime	
$R'O.R_2NP(O)CN$	[R, R', R"... = alkyl]
$R'O.RP(O)F$	
$RP(O)F_2$	
$R'O.RP(O)SCH_2CH_2NR"_2$	
$EtO.RP.OCH_2CH_2NPr^i_2$	
$ClCH_2CH_2SCH_2CH_2Cl$ and homologues	
$ClCH{=}CHAsCl_2$ and homologues	
$Ph_2C(OH)COOQ$	[Q = 3-quinuclidinyl]

Chemicals thus far listed for the *intermediate* control régime	
R-P....	[R = Me, Et or Pr]
$R'_2NP(O)X_2$	[X = hal or OR"; R' and R" = alkyl]
$^iPr_2NCH_2CH_2Z$	[Z = hal, OH or SH]
$AsCl_3$	
$Ph_2C(OH)COOH$	
QOH	[Q = 3-quinuclidinyl]

Chemicals thus far listed for the third control régime (*confidence building*)				
$COCl_2$	CNCl	HCN	Cl_3NO_2	
PCl_3	$POCl_3$			
$(RO)_2POH$	$(RO)_3P$			[R = Me or Et]
SCl_2	S_2Cl_2			

biochemical research, perhaps, or for defence laboratories to develop and test antigas measures against - those small quantities would have to be produced in a special facility, of small production capacity, one per country, continuously monitored by the international inspectorate. Schedule [1] in the annexes to article VI of the rolling text contains the chemicals at present listed for this maximally stringent control régime.

The second control category would be for chemicals having industrial or commercial applications that could not reasonably be curtailed but which are chemicals that could also be used to make highly threatening CW weapons. This control régime would provide assurance of non-diversion by tracking each substance through chemical economies by means of materials accountancy and site inspections, rather in the fashion of the nuclear safeguards operated by the IAEA.[3] All corporations using such chemicals would, therefore, be required to report rather detailed data to control authorities and be required to receive inspectors at the facilities they had declared for the chemicals. Schedule [2] contains the chemicals thus far listed for this control régime; it is at present limited to "key precursor chemicals" only.[4]

The last control category would be for chemicals having much the same "dual purpose" attributes as the second category, but which are far too widely and heavily used for any sort of detailed tracking through an economy to be possible. Production and consumption data would be required from corporations using the chemicals, but that would be for confidence-building purposes rather than for strict verification of non-diversion. Schedule [3] contains the chemicals thus far listed for this control régime.

The hope that is evidently uppermost in the negotiators' minds is that the divide between the second and the third control categories - a divide between chemicals in small and large civil use - is also a divide between the more threatening and the less threatening dual-purpose chemicals. Much of the industry's goodwill towards the treaty depends upon it. Likewise the character of the first divide. Current experience in the Australia Group - the 19 OECD countries that are trying to harmonize their chemical export control policies in the interests of constraining CW proliferation - is not totally encouraging, as Table 23.2 illustrates. In the negotiation on the control lists, and on the associated matter of controlling civil production capacity that has never actually been used to make scheduled chemicals but which could be so used, the merits and demerits of specifying a fourth control schedule have been addressed. That negotiation continues, meshing with negotiation on the "ad hoc" inspection idea and on the most delicate matter of how high the various production/consumption thresholds should be set before the data-reporting requirements come into operation.

It is perhaps not putting it too strongly to observe that, while these latter questions are being resolved, civilian chemical industry has a stranglehold on the whole CW arms control enterprise.

23.4 Notes and References

1 The following 40 countries are members of the CD: the "Western Group", comprising Australia, Belgium, Britain, Canada, France**, Italy, Japan, the Netherlands, the USA** and West Germany; the "Group of Socialist States", comprising Bulgaria, Czechoslovakia, East Germany, Hungary, Mongolia, Poland, Romania and the USSR**: China; and the "Group of 21" comprising Algeria, Argentina, Brazil, Burma*, Cuba, Egypt*, Ethiopia, India, Indonesia, Iran*, Kenya, Mexico, Morocco, Nigeria, Pakistan, Peru, Sri Lanka, Sweden, Venezuela, Yugoslavia and Zaïre. Delegations from a dozen or so other countries have, by invitation, participated in the work of the CD Ad Hoc Committee on Chemical Weapons; Finland, Norway and Spain have been especially active.

Of the known CW-weapon possessor states, which are indicated by a double asterisk (**) in the above listing, Iraq is unrepresented; and unrepresented among the seemingly more probable of the unconfirmed ones (*) are Israel, Libya, North Korea, South Korea, Syria and Taiwan.

Table 23.2. Chemicals subject to national export controls in NATO countries as of March 1988. [a] The conventions used here are as follows: B for blister, N for nerve and P for psychochemical agent; CW 'agents' are made from 'precursors' which are made from 'intermediates' (which are made from 'raw materials'); [b] $MeP(OEt)OCH_2CH_2NPr^i_2$; [c] Not included in the control schedules of the latest draft Chemical Weapons Convention.

Chemical	Single-step product into which the chemical is convertible [a]	Is the chemical subject to Australia-Group export controls in the following countries?															
		NL	US	CA	IT	NO	DK	SP	BR	LX	BE	GE	FR	PO	GR	TU	IS
Thiodiglycol	Blister agent	Yes	Yes	Yes	Yes	Yes	Yes	Yes	Yes	Yes	Yes	Yes	Yes	Yes	Yes	No	No
$POCl_3$	N-intermediate	Yes	Yes	Yes	Yes	Yes	Yes	Yes	Yes	Yes	Yes	Yes	Yes	Yes	Yes	No	No
$MePO(OMe)_2$	N-precursor	Yes	Yes	Yes	Yes	Yes	Yes	Yes	Yes	Yes	Yes	Yes	Yes	Yes	Yes	No	No
$MePOF_2$	Nerve agent	Yes	Yes	Yes	Yes	Yes	Yes	Yes	Yes	Yes	Yes	Yes	Yes	Yes	Yes	No	No
$MePOCl_2$	N-precursor	Yes	Yes	Yes	Yes	Yes	Yes	Yes	Yes	Yes	Yes	Yes	Yes	Yes	Yes	No	No
$(MeO)_2POH$	N-intermediate	Yes	Yes	Yes	Yes	Yes	Yes	Yes	Yes	Yes	Yes	Yes	No	No	No	No	No
PCl_3	N-intermediate	Yes	Yes	Yes	Yes	Yes	Yes	Yes	Yes	Yes	Yes	Yes	Yes	No	No	No	No
$(MeO)_3P$	N-intermediate	Yes	Yes	Yes	Yes	Yes	Yes	Yes	Yes	Yes	Yes	Yes	No	No	No	No	No
KF [c]	N-precursor	No	Yes	Yes	No	No	Yes	Yes	Yes	No	No	No	No	No	No	No	No
$SOCl_2$ [c]	N-intermediate	No	Yes	Yes	Yes	Yes	Yes	No	No	No	No	No	No	No	No	No	No
$^iPr_2NCH_2CH_2Cl$	Nerve agent	Yes	Yes	Yes	No	Yes	No	No	No	No	No	No	No	No	No	No	No
$^iPr_2NCH_2CH_2SH$	Nerve agent	Yes	Yes	Yes	No	Yes	No	No	No	No	No	No	No	No	No	No	No
3-Quinuclidinol	Psycho agent	Yes	Yes	Yes	No	Yes	No	No	No	No	No	No	No	No	No	No	No
3-Piperidinol	Psycho agent	Yes	Yes	Yes	No	Yes	No	No	No	No	No	No	No	No	No	No	No
Me_2NH [c]	N-precursor	No	Yes	No	No	No	Yes	Yes	Yes	No	No	No	No	No	No	No	No
$ClCH_2CH_2CH$ [c]	B-precursor	No	Yes	No	No	No	Yes	Yes	No	No	No	No	No	No	No	No	No
Me_2NHCl [c]	N-precursor	No	Yes	No	No	No	No	Yes	No	No	No	No	No	No	No	No	No
$EtPOF_2$	Nerve agent	Yes	No	No	Yes	No	No	No	No	No	No	No	No	No	No	No	No
$Me_3CCH(Me)OH$	Nerve agent	Yes	No	No	Yes	No	No	No	No	No	No	No	No	No	No	No	No
$EtPOCl_2$	N-precursor	Yes	No	No	Yes	No	No	No	No	No	No	No	No	No	No	No	No
$EtPO(OEt)_2$	N-precursor	Yes	No	No	Yes	No	No	No	No	No	No	No	No	No	No	No	No
$Me_2NPO(OEt)_2$	N-precursor	No	No	No	Yes	No	No	No	No	No	No	No	No	No	No	No	No
HF [c]	Nerve agent	No	No	No	No	No	No	No	Yes	No	No	No	No	No	No	No	No
$(EtO)_2POH$	N-intermediate	Yes	No	No	No	No	No	No	No	No	No	No	No	No	No	No	No
$MePCl_2$	N-intermediate	Yes	No	No	No	No	No	No	No	No	No	No	No	No	No	No	No
$^iPr_2NCH_2CH_2OH$	N-precursor	Yes	No	No	No	No	No	No	No	No	No	No	No	No	No	No	No
$EtPCl_2$	N-intermediate	Yes	No	No	No	No	No	No	No	No	No	No	No	No	No	No	No
Substance QL [b]	Nerve agent	Yes	No	No	No	No	No	No	No	No	No	No	No	No	No	No	No
$(EtO)_3P$	N-intermediate	Yes	No	No	No	No	No	No	No	No	No	No	No	No	No	No	No
$Ph_2C(OH)COOH$	P-precursor	Yes	No	No	No	No	No	No	No	No	No	No	No	No	No	No	No
$MeP(OEt)_2$	N-precursor	Yes	No	No	No	No	No	No	No	No	No	No	No	No	No	No	No
$EtPO(OMe)_2$	N-precursor	Yes	No	No	No	No	No	No	No	No	No	No	No	No	No	No	No
$EtPF_2$	N-precursor	Yes	No	No	No	No	No	No	No	No	No	No	No	No	No	No	No
$MePF_2$	N-precursor	Yes	No	No	No	No	No	No	No	No	No	No	No	No	No	No	No

A discussion of the current state of CW proliferation matters is contained in: J. P. Perry Robinson, Some developments over the past year in the field of chemical warfare armament, a paper for the 13th Workshop of the Pugwash CW Study Group, Geneva, January 1988

2 For a more extended discussion of some of the points raised, see: J. P. Perry Robinson, Verifying a ban on chemical-warfare weapons, Faraday Discussion Paper no. 12, London: The Council for Arms Control, July 1988. The research underlying it was made possible by a grant from the John D and Catherine T MacArthur Foundation

3 See: James F. Keeley, International Atomic Energy Agency safeguards: observations on lessons for verifying a Chemical Weapons Convention, Arms Control Verification Occasional Papers no. 1, Ottawa: Department of External Affairs, Arms Control and Disarmament Division; Bas ter Haar and Piet de Klerk, Verification of non-production: chemical weapons and nuclear weapons compared, Arms Control, vol. 8, no. 3, pp. 197-212, December 1987

4 For an extended discussion of the "precursor" problem based on a review of production methods for the leading CW agents, see: J. P. Perry Robinson, The chemical industry and chemical-warfare disarmament: categorizing chemicals for the purposes of the projected Chemical Weapons Convention, SIPRI Chemical & Biological Warfare Studies, no. 4 , pp. 55-104, 1986

Contributors

Dr. Jürgen Altmann

Physicist, Department of Physics and Astronomy, Ruhr-Universität Bochum, FRG. Recent publications on laser weapons and anti-tactical ballistic missile defenses. Works in research project "New technical means for cooperative verification in Europe"

Dr. Alexei Arbatov

Institute of World Economy and International Relations (IMEMO), USSR Academy of Sciences, Moscow. Head of Department of Disarmament and International Security, Head Author of the Disarmament and Security Yearbooks of IMEMO (Moscow: APN)

Dr. Roger A. Clark

Geophysicist, University of Leeds, UK

F. Ronald Cleminson

Colonel, Head of Verification Research Unit, Arms Control and Disarmament Division, Department of External Affairs, Ontario, Canada. Coauthor of: "Compendium of Arms Control Verification Proposals"

Dr. Gloria Duffy

Political Scientist, President of Global Outlook, Palo Alto CA. Recent book: Compliance and the Future of Arms Control (Stanford CA: Stanford University and Global Outlook, 1988)

Dr. Steven Fetter

Physicist, School of Public Affairs, University of Maryland, College Park MD, USA. Recent book: Toward a Comprehensive Test Ban (Cambridge MA: Ballinger, 1988)

Dr. David A. V. Fischer

Cambridge, UK. Formerly Assistant Director General for External Relations at International Atomic Energy Agency

Dr. Bernd A. Goetze

Colonel, Canadian NATO Delegation, Brussels, Belgium. Recent book: Security in Europe: A Crisis of Confidence (New York: Prager, 1984)

Manfred Graczynski

Captain, Ministry of National Defense, Berlin, GDR. Member of GDR delegations at MBFR negotiations, Stockholm Conference on Confidence and Security-Building Measures and Disarmament in Europe and of Vienna Follow-up Meeting of the CSCE.

Eng. Klaus Jacob

Lieutenant Colonel, Engineer, Arms Control Agency of the Western European Union, Paris

Viktor Koltunov

Colonel, USSR General Staff, Moscow, USSR. Member of USSR delegation to Geneva Strategic Arms and Space Negotiations.

Dr. Alfred Lecocq

Physicist, Collectif des Scientifiques pour le Désarmement Nucléaire, Gometz Le Chatel, France

Dr. Jeremy Leggett

Geophysicist, School of Mines, Imperial College of Science and Technology, London, UK. Director of Verification Technology Information Center, London

Dr. Harald Müller

Political Scientist, Peace Research Institute Frankfurt

Julian P. Perry Robinson

Science Policy Research Unit, University of Sussex, Brighton, UK. Book and journal publications on chemical weapons and chemical disarmament

John Pike

Federation of American Scientists, Washington DC, USA. Director of Space Policy of FAS

Dr. Oleg F. Prilutskii

Physicist, Cand. Sc., Institute of Space Research, USSR Academy of Sciences, Moscow, USSR. Coauthor of: Weaponry in Space - The Dilemma of Security (Moscow: Mir, 1986)

Roland Reimers

Physicist, Institute for Peace Research and International Security, Hamburg, FRG

Dr. Stanislav N. Rodionov

Physicist, Dr., Cand. Sc., Institute of Space Research, USSR Academy of Sciences, Moscow, USSR. Coauthor of: Weaponry in Space - The Dilemma of Security (Moscow: Mir, 1986)

Prof. Joseph Rotblat

Physicist, University of London, UK. President, Pugwash Conferences on Science and World Affairs. Research in nuclear physics and radiation biology. Book and journal publications on science, scientists, the arms race and disarmament

Dr. Heinrich Siegmann

Computer Scientist/Political Scientist, Dr., European Center for International Security, Starnberg, FRG

Nikita P. Smidovich

USSR Ministry of Foreign Affairs, Moscow, USSR. Member of USSR delegation to Geneva Conference on Disarmament

Dr. Robert A. Summers

U.S. Arms Control and Disarmament Agency, Washington DC, USA. Chief of Division of Verification, Bureau of Verification and Intelligence

Dr. Theodore B. Taylor

Physicist, West Clarksville NY, USA. Formerly at Los Alamos National Laboratory, USA, developing nuclear weapons

Dr. Valerie Thomas

Physicist, Dr., Center for Energy and Environmental Studies, Princeton University, Princeton, USA

Abbreviations

ABL	Armored Box Launcher
ABM	Anti-Ballistic Missile System
AIAA	American Institute of Astronautics and Aeronautics
ALCM	Air-Launched Cruise Missile
AOA	Airborne Optical Adjunct
APC	Armoured Personnel Carrier
ASAT	Anti-Satellite Weapon/System
ASROC	Anti-Submarine Rocket
BM	Battle Management
BSTS	Boost Surveillance and Tracking System
BSVRP	British-USSR Seismic Verification Research Project
CBM	Confidence-Building Measure
CCD	Charge-Coupled Device
CD	Conference on Disarmament
CDE	Conference on Confidence-Building Measures and Disarmament in Europe
CFE	Negotiations on Conventional Armed Forces in Europe
CIA	Central Intelligence Agency
CID	Charge Injection Device
CLS	Capsule Launch System
CORRTEX	Continuous Reflectometry for Radius versus Time Experiment
CS	Conventional Stability
CSBM	Confidence and Security-Building Measure
CSCE	Conference on Security and Cooperation in Europe
CSSR	Czechoslovakian Socialist Republic
CTB	Comprehensive Test Ban
CTBT	Comprehensive Test Ban Treaty
CV	Carrier Vehicle
CW	Chemical Weapon, Chemical Warfare
CWC	Chemical Weapons Convention
C^3	Command, Control and Communications
C^3I	Command, Control, Communications and Intelligence
DA	Deployment Area
DARPA	Defense Advanced Research Projects Agency
DOD	Department of Defense
DOE	Department of Energy
ERIS	Exoatmospheric Reentry Vehicle Interception System
ERS	ESA Remote-Sensing Satellite
FBR	Fast Breeder Reactor
FLIR	Forward-Looking Infrared
FRG	Federal Republic of Germany
FUM	Follow-UpMeeting
FY	Fiscal Year

GDR	German Democratic Republic
GSETT	Group of Seismic Experts Technical Test
GSTS	Ground Surveillance and Tracking System
HEU	Highly Enriched Uranium
HWR	Heavy Water Reactor
IAEA	International Atomic Energy Agency
ICBM	Intercontinental Ballistic Missile
IFSH	Institut für Friedensforschung und Sicherheitspolitik an der Universität Hamburg
INF	Intermediate-Range Nuclear Forces (Treaty)
INFCIRC	Information Circular
IR	Infrared
IRM	Intermediate-Range Missile
JSTARS	Joint Surveillance/Target Attack Radar System
JVE	Joint Verification Experiment
LWR	Light Water Reactor
LYTTB	Low Yield Threshold Test Ban
LYTTBT	Low Yield Threshold Test Ban Treaty
MBFR	Mutual Balanced Force Reductions
MBT	Main Battle Tank
MHD	Magneto-Hydrodynamic System
MIRACL	Mid-Infrared Advanced Chemical Laser
MOU	Memorandum of Understanding
MSR	Molten Salt Reactor
MUF	Material Unaccounted For
NASA	National Air and Space Agency
NATO	North Atlantic Treaty Organization
NNA	Neutral and Non-Aligned
NNWS	Non-Nuclear Weapon State
NPT	Non-Proliferation Treaty
NRDC	Natural Resources Defense Council
NTM	National Technical Means of Verification
NWS	Nuclear Weapon State
OECD	Organization for Economic Cooperation and Development
OSI	On-Site Inspection
OSIA	On-Site Inspection Agency
OTA	Office of Technology Assessment
PBV	Post-Boost Vehicle
PGM	Precision-Guided Munition
PNE	Peaceful Nuclear Explosion (Treaty)
PPM	Perimeter-Portal Monitoring
PRIF	Peace Research Institute Frankfurt
RMS	Root-Mean-Square
RV	Reentry Vehicle
SALT	Strategic Arms Limitation Talks
SAR	Synthetic Aperture Radar

SATKA	Surveillance, Acquisition, Tracking and Kill Assessment
SBI	Space-Based Interceptor
SDI	Strategic Defense Initiative
SIPRI	Stockholm International Peace Research Institute
SIRTF	Space Infrared Telescope Facility
SLCM	Sea-Launched Cruise Missile
SNR	Signal-to-Noise Ratio
SPIE	Society of Photo-Optical Instrumentation Engineers
SSTS	Space Surveillance and Tracking System
START	Strategic Arms Reduction Talks
SVC	Special Verification Commission
TASM	Tomahawk Anti-Ship Missile
TLI	Treaty-Limited Item
TNT	Trinitrotoluene
TTBT	Threshold Test Ban Treaty
TV	Television
TW	Throw-Weight
UK	United Kingdom
UN	United Nations
US	United States
USA	United States of America
USSR	Union of Soviet Socialist Republics
UTC	Universal Time Code
VLS	Vertical Launching System
WEU	Western European Union
WGPu	Weapons-Grade Plutonium
WP	Warsaw Pact
WTO	Warsaw Treaty Organization

Subject Index

FSC
www.fsc.org
MIX
Papier aus verantwortungsvollen Quellen
Paper from responsible sources
FSC® C105338

If you have any concerns about our products,
you can contact us on
ProductSafety@springernature.com

In case Publisher is established outside the EU,
the EU authorized representative is:
Springer Nature Customer Service Center GmbH
Europaplatz 3, 69115 Heidelberg, Germany

Printed by Libri Plureos GmbH
in Hamburg, Germany